THE COMPLETE MINER

Production processes using machines and

Equipments to extract Tin and Columbite ores

PURPOSE

1. It is a guide for unskilled artisan miners, prospective tin shed operators and prospective tin and columbite mining investors to acquiring technical skills and knowledge of production process and operations of mining equipments and machines for processing tin and columbite.

2. It serves as a resource base for capacity building to unskilled artisan miners, prospective tin shed Operators and prospective tin and columbite investors to be taught and trained on how to use machines and equipments to extract and process Tin, Columbite, Tantalite and other minerals.

3. It highlights broad insight into the Types, Uses and Principles of operation of major machines and equipments in the tin shed and their areas of

application in the production of tin and columbite and other minerals for optimum performance of unskilled artisans miners, prospective tin shed operators and prospective investors in the tin and columbite mining sub-sector.

4. It is a resourceful material for Mining engineering students as well as Geology and Mining students to acquire additional knowledge which is relevant to their field of study.

5. It serves as a material rich in historical facts about Nigeria's past glory and dominance in the Tin and Columbite solid minerals sub- sector and therefore worthy of archiving by all the industry players.

TABLE OF CONTENT

INTRODUCTION

PRE-PRODUCTION STAGE

[1] **Pre-Machine Processes**

(A) **Sieving**

(i) Sieving Machine

(ii) Wooden/Steel Hand held sieve

(iii) Large Double Handle Sieve

(B) **Grinding and Crushing**

(i) Grinding using Mortar and Pestle

(ii) Crusher Machine

(C) **Jigging**

(i) Machine or Shaking Table Jigging

(ii) Hand or Manual Jigging

(D) **Washing and Boxing**

(i) Stream washing using calabash

(ii) Washing using Jirgi

(iii) Washing using Boxing machine or Wilbi

(E) **Drying**

(i) Normal floor & Rake drying

(ii) Fire plate drying

PRODUCTION STAGE

[2] **Machine Processes**

Gravity Separation Using air

(A) **Air Floating Machine**

(i) Big Air floating Machine

(ii) Small Air Floating Machine

(iii) Air Floating Techniques

(a) Free Floating technique

(b) Hand Brush Floating technique

Magnetic Separation using EMF effect

(B) **Magnetic Separator Machine**

(i) BMF Magnetic Separator

(ii) Rapid Magnetic Separator

(iii) Single Disc Magnetic Separator

(iv) Types of Separator Machine Dressing

(a) Passing

(b) Running

(c) Re-Run

POST-PRODUCTION STAGE

[3] **Post-Machine Processes**

(A) **Mixing and Taking Samples**

(i) Hand Mixing

(ii) Shovel Mixing

(B) **Acid and Powder Burning** (for Tin & Columbite)

(i) Chemical Acid

(ii) The Powder

(iii) The Burning Process

(C) **Burreting or Volumetric Analysis** (for Columbite and sometimes Tin)

(i) The Burreting Tube

(ii) The Tube Holder Stand

(iii) Plastic Cup

(iv) The Scale Balance

(a) Obsolete Justice Scale Balance

(b) Modern Spring Balance:

(v) The Burreting Process

(D) **Analysis** (Columbite, Tantalite & Tin)

(i) Introduction

(ii) Analysis Equipments/Machines

(a) Portable Handheld XRF Analyzers;

(b) Bench top bulk analysis XRF:

(iii) Analysis Process and Presentation

(a) Offsite Analysis

(b) On-site Analysis;

(E) **Other Post Production Tests** (Wolfromite & Lead)

(i) Wolfromite Test

(ii) Lead Test

(F) **Weighing**

(i) The weighing Scale

(ii) The Steel Post;

(iii) The weighing Bucket

(iv) Standard techniques for weighing Tin and Columbite

(a) Handle and Hook Technique

(b) Double Handle & side technique

(G) **Bagging**

[4] GRADE BOOSTING TECHNIQUES

(A) **Roasting**
(i) Roasting of low grade columbite ore
(ii) Roasting of Monoxide ore

(B) **Combinations/Blends**
Boosting Combinations
(i) Boosting low grade columbite ore with tin ore
(ii) Boosting low grade columbite ore with wolframite ore
(iii) Boosting low grade columbite ore with roasted monoxide
(iv) Boosting low grade columbite ore with Lead ore
(v) Boosting low grade columbite ore with Low grade Tantalite ore

[5] FINAL PRODUCT
(A) **Grading**
(B) **Storing**
(i) Internal storage
(ii) External storage

(a) Raw mineral ore Storage

(b) Processed mineral product Storage

(C) **Transportation and recommended vehicles**

INTRODUCTION:

Geologically, Tin (Sn), Niobium (Nb), and Tantalum (Ta) are found in the late stage magmatic products, such as pegmatite and high-temperature veins. Common minerals of these elements are

Cassiterite (SnO_2)

Columbite (FeMn) Nb_2O_6

Tantalite (FeMn) $Ta2O6$

Wolframite (FeMn) $WO4$

Scheelite ($CaWO_4$)

Tantalite is commonly found with columbite (FeMn) Nb_2O_6. This close association is called columbite-tantalite or Coltan (Ta-Nb). These dense minerals can be reworked, eroded, transported by water, and accumulated due to gravity separation during sedimentary processes. Such deposits are called placer deposits, which are common deposits for the mining of Nb, Sn and Ta in some countries.

In application, Tin (Sn) has been used for thousands of years in numerous applications due to its low fusion point. Since 2006 when lead (Pb) was forbidden in welding products in some countries, the demand for Sn has increased dramatically for its use in solders.

Niobium (Nb) is used in the steel industry, as well as in super-alloys and superconductors. Because Tantalum (Ta) is a very robust metal with a high fusion point, it is used in missiles, airplanes, and throughout the nuclear industry. Niobium (Nb), Tin (Sn) and Tantalum (Ta) are mainly used in high-end technology products.

PRE- PRODUCTION STAGE:

The pre-production Stage is so named to specify all activities carried out at the tin shed prior to the involvement of the actual extraction machine(s). This is one of the most important phase in the value chain system of tin and columbite production because it involves activities whose success are conditional to the success of subsequent operation. Activities in this stage must be undertaken to pave way for future activities to take place.

(1)Pre-Machine Process:

There are two types of Tin and Columbite ores dug from the ground or mined from paddocks; these are pure ore which is 95% rich and impure ore which comes in various percentage range of purity. The pure ore has about 5% or fewer impurities in the ore. The amount of processing to be carried out on the ore depends on the quantity of impurities in it. The pure dug mineral ore is usually Black or Blue Black in color like magnetic tin and it is mostly coarse aggregate in structure. The pure excavated ore could be as large as 3-5 cubic inches in size until it is broken down into smaller aggregate sizes through crushing machines or mortars.

The second type of ore which is impure has high quantities of impurities and is loosely formed. This type of ore results from the excavation of the immediate and surrounding areas where the pure parent ore was dug previously. Another source of the impure ore is the underground dump being left fallow for a long period of time. The impure ore which is also referred to as "Tailings" is equally popular amongst miners as "Material" which is a miners jargon. Tailings come in diverse natural colors which are influenced by the amount of impurities in them, such as Black, Blue-Black, Brown and Reddish-Brown and Milky-Sand colors.

A normal impure Tailings material or mineral ore,

contains many inherent substances such as iron, columbite, tin, sand, monoxide, zircon while some materials even contains little fragments of wolfromite and tantalite. All these could be constituents of a Tailing material whereby each one actually have their economic value. Amongst the constituent substance of a Tailing material listed above, tin and columbite are the most naturally widespread and having large reserve deposits within the greater Jos Bukuru axis. They also have stable global markets since the 1930s with periodic rise and fall in prices. Monoxide is a yellowish-Red colored substance and it is a very heavy substance. Monoxide usually clings closer to Columbite than it does tin. Although in any case it has to be separated from the two in other to leave the mineral elements in their pure form and weight.

Tin is a Black or sometimes Dark-Brown color mineral

substance that is extracted from pure or impure tailing materials. Tin is relatively heavier than Columbite in weight. In its pure refined form, Columbite is a glittering and shiny Blue-Black mineral with a rectangular diamond shape and looks very attractive to the eyes. Iron is a Black or Bluish-Black mineral substance contained in a Tailing material or ore. It is virtually the lightest mineral substance in an ore or Tailing. It is the most unwanted amongst the constituent elements of a Tailing material which has no any economic value. It constitutes a burden to the miner in that it must be separated from the others. The presence of iron in any material ore makes the ore to be lighter in weight. The sooner iron is separated or removed from any Tailing material; the material will gain its pure form and weight. Zircon is a brown, lightweight substance found along with the other substances in a Tailing material. It is found in little quantity in most mineral ores, as compared to iron. Zircon is mostly extracted at the front bell or over belt of Magnetic Separator Machines along with some tiny whitish sand particles. It has a rubber-like appearance. When it comes along with tin or columbite it lowers the weight of these minerals and therefore their quality. When the Tailing material is brought out of storage or pit, the first line of action in the processing center is sieving.

(A) Sieving:

The main reason for sieving Tailing material is to separate the soft or fine particles from the coarse particles. This is to prepare the material for air floating or running in some cases in a Magnetic Separator. The two main machines cannot perform optimally if tailing or pure tin and columbite ores are not sieved. This is because the machine can only contact the coarse aggregates first due to their height and size while leaving the fine aggregates to flow unattended to. When sieving is not carried out, the bigger particles which are the coarse aggregates will first make contact with the electro-magnetic discs of Separators and it will be processed accordingly while the soft particles which are the fine aggregates will not be processed. For Air Floats Machines, when Tailing materials are not sieved, coarse aggregates will follow number 1 and 2 buckets even when they are not good grades while the good grades amongst the fine aggregates will flow to the low grade buckets of 3 to 5 even when they are actually good quality grades.

This is while sieving, though looking ordinary but it's one of the most important process in tin and columbite production, and that is why it precedes other processes.

Most at times, it is discovered that in some materials, the coarse particles are richer in tin and when sieved, the pure tin can be separated from the other impurities easily without undergoing some lengthy processing procedures, thereby necessitates sieving.

There are three main types of sieves; Sieving Machine, Wooden Hand held sieve and Large Double Handle Sieve.

(i) **The Sieving Machine:** The most popular sieve machine used in filtering tin and columbite ore is the Ark trademark machine. It is an electro-mechanical machine which works on the principle of vibration. This sieving machine is completely round in shape and has quadruple leg stand with some having small rollers in each leg for moving the sieve from place to place. The sieve machine has a height of 1.3-1.5m; it has a single cylindrical trough for pouring the Tailing material to be sieved inside having a diameter of 0.7 – 0.9m and trough height of 0.75m. Half way inside the trough, that is 0.375m from the covered top is the main sieve filter with many holes or perforations.

The sieve has three clip hooks around it used to hold it firmly in place and also for adjustment during cleaning maintenance or removal during change of a different sieve size. Several sieve sizes can be used to suite different material and to obtain different mineral grades. The top has a hinged cover which can be opened to fill Tailing material and closed during the filtering process. The compartment immediately below the sieve has two mechanical divisions opposite each other and linked to exit funnels to the outside. This funnel discharges the two resulting sieved materials to collecting buckets. One of the buckets collects the fine sieved material while the other bucket collects the coarse top retained material.

(ii) **Wooden/Steel Hand held Sieve:** The second type of sieve used in the processing of tin and columbite is the most popular and most common found in all tin sheds unlike the sieving machine which is only available in some large mining companies who can afford it. The Wooden sieve is made from very hard timber to withstand pressure from constant usage and mishandling. The Wooden hand-held sieve is constructed permanently with a particular sieve size unlike the sieving machine whose sieve can be removed and changed to another size.

As the name implies, this type of sieve is held with both hands and vibrated to and fro. In the process of sieving, the finer particles penetrate the little holes or perforations on the sieve net and fall below. The bigger particles which could not pass through the sieve mesh are retained on the sieve surface. The materials used to make the rectangular or square sieve frame or body are mostly fabricated with wood, reinforced plastic or steel to give it super strength because of the constant pressure of wear and tear as explained earlier. While the perforated mesh itself as the underside filter is made from steel wire mesh, polyvinyl strands or canvas. The sieve comes in different mesh and body sizes. The mesh perforation sizes include sizes 5, 10, 20, 30, 40, 50, and 60, to 100. The most common sieve size in constant use in the tin shed for processing of tin and columbite are sizes 20, 40 and 60.

(iii) **Large Double Handle Sieve:** This sieve is fabricated for heavy duty purposes. Some are 1.5m x 0.8m rectangle while others could be as large as 2m long by 0.9m wide rectangle. The four handles are 0.6m long each. The sieve is fabricated from strong timber or steel body frame while the four handles are made from the same material. The mesh net is made from steel wire strands or from reinforced polyvinyl material. This sieve is constructed permanently with a specific mesh size and cannot be changed like that of the sieving machine. This sieve is made large enough and can only be used by two people facing each other. This type of sieve is principally used to filter raw Tailing material that was dug or sourced from old dump and needs to be separated from stones, pebbles, broken bottles, grasses and tuffs and many other impurities. The process is carried out by three people; one person uses a shovel to consistently feed the large sieve with tailing material after the top retaining coarse aggregates are thrown-off. The remaining two people will carry out the main sieving process by grabbing the handles using both hands by each person at opposite sides while standing up and moving the sieve to and fro in unison, until the fine particles goes through the sieve to the ground while the larger particles which are retained on top of the sieve are thrown away.

(B) **Grinding and Crushing:**

The next important pre-machine process is grinding and crushing. The principal reason for grinding is to convert the coarse aggregates into fine powder. When this is done, all impurities like zircon and monoxide which stick closely to pure tin or columbite ore and refused to separate from it is crushed to very fine powder and it is blown off into the air through the use of Matankadi(local circular trough) or through Air float Machine thereby freeing the minerals into their pure form.

(i) **Grinding Using Mortar & Pestle:** Grinding is done manually with hand using a cast iron mortar and pestle. It is done vertically by lifting the pestle up and dropping it down with substantial amount of force to land on the particles continuously and deliver an impact capable of crushing the particles and converting it into fine powder form.

(ii) **Crusher Machine:** Crushing however is done on a machine called the Crusher. The crushing machine could be powered by electricity or diesel. This machine looks like a small sitting lister engine. It is a composite metallic machine which operates on the principle of power momentum; it crushes coarse aggregates through applied force by action of gears and metallic teeth. The crushing machine is characterized by strong metallic teeth or gears that run in opposite direction to one another. The Crusher tooth is operated through selection gear handle provided at the side of the machine to increase or decrease the power rating of the machine. This machine crushes different grades of ores from coarse to bigger size aggregates up to 10 to 12 inches in volume. In both cases, this process is carried out to break down the hard solid ore so that it will make it easier for the impurities to be separated from the important mineral in its loose form.

Another reason for crushing is to disintegrate those big aggregates which cannot be otherwise grinded through manual action of Mortar and Pestle. After crushing, the relative fine mineral particles would then be sieved or Air floated depending on the amount of impurities in the ore.

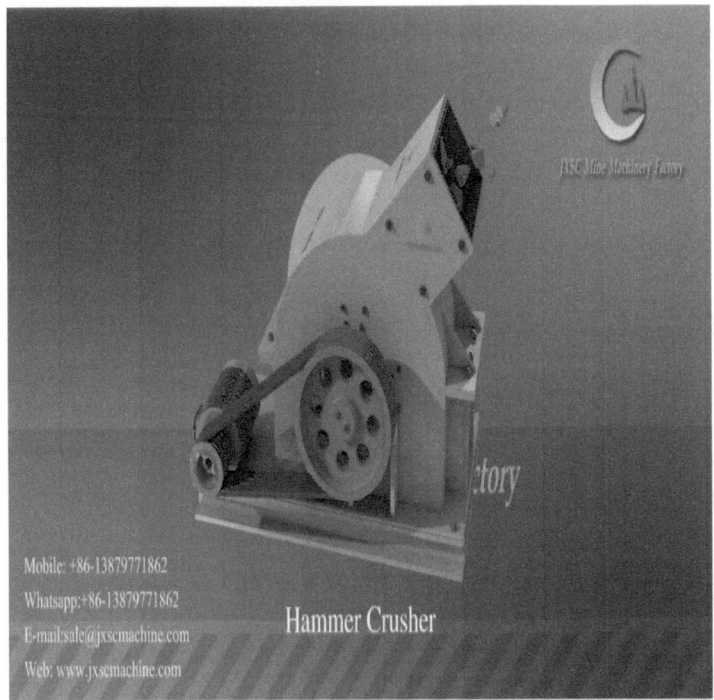

(C) **Jigging:**

This process is used for the extraction of, particularly Tin ore through the combine action of water and manual or mechanical vibration depending on the type of Jigging process undertaken. The process is also used for separating important ore from among other impurities.

The material particle to undergo this process should be coarse or should be the particle size retained on a sieve. Jigging process is strictly water based or rather it is strongly associated with water. There are two types of Jigging. There is the machine or table jigging and the hand or manual jigging.

(i) **Machine or Shaking Table Jigging** is done using a wide table-like platform machine which oscillates in a constant defined periodic motion. The machine is 3.5 to 4 feets high from the ground with vertical standing legs. It has a length ranging from 15 to 20 feets long and the width is about half the length. It is only six (6) inches thick. It has water pipes running along its length. The table Jig is powered by electricity or diesel with its motor and belt drive.

Working principle; the working principle of Shaking

Table is to use the combined action of the specific gravity difference of sorted minerals, alternating movement of bed surface, and transverse oblique water flow and riffle (or notch groove), to allow loose layering of ores on the bed surface and fan-shaped zoning. Then different products can be produced. The following Major parts of the shaking table plays significant role in its effective functionality viz;-

Head motion; Shaking tables are furnished with a totally enclosed self-oiling head motion of heavy cast iron to contain an oil reservoir for perfect splash lubrication. This feature protects the moving parts, reducing operating and maintenance cost to a minimum.

Deck; The decks are constructed of steel framework and covered 16mm fiberglass with corundum, this cover its strength reaches as high as 70% of steel, the features of this design are stronger, wear-resistant, corrosion-resistant, and no distortion, it lengthens the working-lige of machine and thus allows machine perform perfectly on different mineral and different weather condition.

Feed and water box; A wooden feed distributing box with hopper and a long water box are attached to the iron of the deck, giving a very even distribution of feed and water. The machine works on the concept of vibration mechanism aided by the force of water, the flowing water on the Jig table separates the lighter particles from the heavier ones by forcing the lighter ones backwards and the heavier ones continue to move by virtue of their weight. The constant vibration of the table makes the already water separated particles to now juxtapose and move their separate ways along the table to different collection pockets.

The system is a continuous one as long as there is

availability of power and water. The resulting mineral ores from the heavy pocket is pure Tin which can be sold right away, while the lighter minerals from the other pocket contains Columbite, iron, sand and other impurities which will have to undergo other production processes to extract the Columbite.

The Shaker Table is also widely used in separation for Gold, Silver, Tin, Columbium, Tantalum, Titanium, Barium, Tungsten, Iron, Chrome, Manganese, Zircon, Lead, Zinc, Mercury, Copper, Aluminum, etc. The shaker effective recycling granularity scope is 2-0.22 millimeters.

(ii) **Hand or Manual Jigging:**

is the opposite version of the machine jigging process. The miner must be experienced to undertake hand Jigging, this is why only 1 in 12 miners know how to Jig manually. Here, the miner will have in front of him a large bowl which may have diameter range of between 0.5 to 1m to allow enough space for motion and maneuvering. The miner will equip himself with a strong sieve (called "Lariya" in Hausa language). This sieve is quite different from the sieve for normal tailing material sieving. The sieve for Jigging has larger perforated holes and very strong wooden handle made of Mahogany. As stated earlier Jigging is made for coarse aggregates called "Gwaragwara" in Hausa language. This, the miner will fetch a sizable quantity and pour into the sieve. Before commencing, the sieve is inclined at an angle less than 90 degrees in front of the miner with half of the sieve sinked into the bowl. The front side of the sieve is positioned facing upwards while the back side facing the miner is dropped into the water in an inclined position inside the bowl. The miner wraps both his hands around the sieve with each hand grabbing each side of the sieve.

The actual Jigging is done by an initiated motion to and fro in a constant crescendo and at the same time applying force to the sieve through the water resistance in the bowl.

(D) **Washing and Boxing:**

Tailing Washing is one of the most important and relevant activities in the pre-machine process. Most Tailings or material when brought to the tin shed will be taken straight to the washing place. Although there is no clear cut rule as to the distance or dimension of a washing space. But The washing space is always located 8 to 12m away from the tin shed; this is because of the high volume of water usage during Tailing washing which can lead to flooding into the tin shed if close enough or mosquitoes breeding ground, the bigger the tin shed the wider the washing space and the more the washing equipments.

In some tin sheds, the washing space is provided under a makeshift canopy or open wooden tent for protection of the workers from the effect of rain and direct sunlight. Washing of tailing material in fact leads to a rise of the material grade up to ten steps and even above. For instance when a Columbite is Buretted and is found to be 20.5 grades, it falls into a low grade category. When this columbite is washed thoroughly, and re-Buretted again, the grade can shoot up or rise to be 19.4, which is a good grade next to the highest grade.

The main reason for washing Tailing material is to do away with all impurities constituent in the Tailing material thereby raising its grade. Some dumped ores left for years contained all manners of impurities in them like broken glass, stones, iron, and vegetative plants and so on. In other to eliminate such impurities, the material must first undergo careful washing. Any Tailing material that contains the earlier mentioned impurities must not be taken directly to the machine because it will block the machine outlet which is the lower end of the big trough containing a narrow funnel-like opening linked to the big trough where the Tailing material is poured.

There are three types of Tailing material washing in tin and columbite processing. The first type is the stream washing using only calabash while the second type is the most popular using an open top box wooden construction called "Jirgi" and the third type is done using the Boxing machine or "Wilbi".

(i) **Stream washing using calabash** is done in shallow, slow moving stream. This type of washing is mainly done while standing or prostrating. It is a sole activity where an individual does everything. The essence of this washing is to acquire and isolate tin and columbite ores from running water while fresh. It is a known fact that running water forces most heavy and light mineral ores to flow along the stream in the direction of flow of the water leaving in its trail black trace of the ore. In this case it transports many important minerals and precious stones, tin and columbite inclusive. This assertion also means that tin and columbite along with other precious stones can be found and mined both on the earth surface, under the earth strata and in moving streams.

The process always begins with the miner using a shovel to create an embankment with the sand ore particles flowing through the stream in that direction. The fast moving sand ore particles are not mere sand particles but a combination of sand and mineral ore containing other impurities. After creating the embankment, the miner then uses his Calabash and water from the stream to wash the mineral ore which is part of the embankment. His effort will be aided by the activity of the running water on the sand embankment trying to force its way through, thereby splashing water and forcing the lighter sand away and leaving the relatively heavier minerals where they are. Subsequently, the miner will use his calabash to wash the mineral ore after he poured a sizable quantity into the calabash. The water will not be filled to the brim and he uses both hands to grasp around the circumference of the calabash to spin the calabash continuously in circles and half circles until he separates the sand and little pebbles from the mineral ore. This process does not segregate the pure mineral (tin and columbite) from the unwanted ores and impurities (including iron, monoxide and zircon), the machines does that.

THE COMPLETE MINER

(ii) **Washing using Jirgi** is done around the tin shed at the designated washing area. This type of mineral Tailing washing is carried out while sitting down and must be assisted by one or two people. While the washer sits down, the others work while standing and they being the apprentice do the most manual function of serving the Tailing material (using shovel for direct washing or using cut sack for boxed materials), constantly filling water into the short drum and evacuating the waste from the back head pan. The main washing equipment called Jirgi (meaning train or plane in Hausa language) is constructed from very strong timber and made in an inclined shape with the top higher than the rest of the body. It measures 2.5m x 0.4m x 0.2m at the top and 0.05m at the lowest tail end. It slopes down gently until it reaches its lowest end.

Dimensionally, it is constructed in the shape of an open

top, gentle sloping box. The top has a flat square area filled with pebbles and sand to allow water to rest there and reduce the velocity of running water while at the same time enabling a wider spread of water to cover the width of the trough. The remaining body of the Jirgi which is gentle sloping is open, smooth and free fall for running water and Tailing material to move. A short half drum is placed on top of some pre-arranged sand filled bags above the Jirgi to supply water to the Jirgi. A small hole of 20mm is perforated at the underside of one side of the circular drum to allow water to fall onto the Jirgi. A control is placed on the hole to control the flow of water. The washer used the steel control wrapped with clothe to control the opening, speed and closing of water.

(iii) **Washing using Boxing machine or Wilbi:** the Wilbi or boxing machine is all steel manually operated equipment. It has a tall steel cylindrical water drum which measures 6feets high and 2feets Diameter and a square steel open top box which measures 2.2feets square and 4feets high. The cylindrical drum and the square box are screwed the same flat steel base which itself is 5-6inches thick and has hollow interior. The Wilbi has a long steel handle which is connected to a circular base cover plate inside the cylindrical drum. The bottom of the steel square box is covered with a perforated sieve with tiny holes which allows the passage of water. The principle of operation of the boxing machine or Wilbi is the forceful impact of water on the Tailing material inside the square box. For a start, water will be filled up into the cylindrical steel drum by an apprentice and continuously do so even when the boxing process is going on. The miner will first of all fill the square box with Tailing material before going to close the bottom water exit of the Wilbi with a cover then he will apply certain amount of pressure on the steel handle until he overcomes the opposing resisting forces of water on the cover base plate to open the base and the sudden pull will send water forcefully gushing out for space and rushed through the hollow bottom space and force itself up through the perforated holes in the sieve and up into the Tailing material soaking and lifting some of the materials high above up to 1.6 feet above the box. This action makes the whole dissolved and water filled Tailing material to boil and utter some gurgling sound. At this juncture, the miner will insert or immerse one of his bare hands inside the box while still holding the steel handle with the other hand, until he touches the Tailing material, moving his bare hand all around inside the box to the bottom of the box and even touching the sieve. This action is meant to ensure that proper dissolution and mixing of the material took place.

When set, the miner drops the steel handle which automatically drives back the cover plate and shuts off the water source, he then opens the base cover to allow all the water inside the hollow bottom to escape thereby making the water inside the square box to also escape and leave the Tailing material inside the box to settle and dry. The resultant boxed material settles into a layered strata where the top most material is cut and thrown away while the bottom material is coarse aggregate in size and very rich in tin ore called "Gwaragwara" which will further undergo hand or machine Jigging to extract the final Tin.

(E) **Drying:**

In Tin and Columbite processing, Drying is very important, although it may not sound so. All ores or

Tailing materials must be in its dry form to be able to undergo Machine process with the exception of Jigging process which will not work without water. Tailing materials, whether dug, washed or jigged must first be dried to prepare it for further processing. Every tin shed must have a wide open space, and well screeded drying area to be located not far from the washing area. Sometimes, the drying space is not enough because of the quantity of materials that needed to be dried. When there is so much material to be dried, people will have to resort to rationing of available space so that every miner can dry his or her material at his allocated time and day. The following includes the types of drying in tin and columbite production:

(i) **Normal floor & Rake drying**: This type of drying is done in an open space provided there is a pre-constructed floor for it. It is the most common type of drying in the processing of tin and columbite ores. The process is very simple and does not require any expertise to undertake in it, anybody can do it. It only requires shovel, Rake and Bags. Since the drying is done in an open space, it means the ultraviolet sun rays from the Sun and air dries the Tailing material. The first action is to spread the wet Tailing material lightly and widely on the floor using a shovel. Next, the miner uses a Rake to spread the Tailing material lightly and he continuously periodically rake the material to allow air to get access to the wet materials below and to up- turn the materials below to expose them to direct sunlight.

When the Tailing material is totally dry, the miner will use dry bags to evacuate it straight to the store for safe keeping or straight into the tin shed to be positioned for dressing. Drying Tailing material is a herculean task during the rainy season, especially when there is large quantity of material to be dried and also there was a need to dress and supply such material in time. In rainy seasons, there use to be so many drying contract jobs available and so many miners competing for drying space. When a spread Tailing material is dry, the miner uses a strong broom to sweet them together closely, gather them in various piles and then use dry bags to pack them, evacuate and then spread another if there is.

(ii) **Fire Plate Drying:** This type of drying is mostly carried out by individuals and Directors that owned their property. This type of drying is mostly carried out during rainy season and also where there is an urgent need to cover an allocated supply demand. A space is provided for this purpose usually at the back of a tin shed or property. Four (4) large boulders arranged in a rectangular or square order constitute the main platform legs stand while a wide sheet metal of various width and gauge is used as the drying platform. Logs of wood are used to generate fire as the main source of heat.

The metal sheet which is the drying platform will soon absorb and get heated after some time and thereafter the Tailing material will be poured on it using shovel and then spread with rake. It takes just a little time to dry on the metal plate as compared to other medium of drying. The challenge with this type of drying is that it needs constant raking otherwise the material will become very hard and stiff like stone as a result of heat and will be difficult to be separated.

PRODUCTION STAGE:

The production Stage, as the name implies refer to all the activities that are done to actually extract the pure ores from tailing materials using the main extraction machines available. This stage made use of mostly experienced personnel or miners more than the previous stage.

[2] **Machine Processes:**

This process begins when the material is taken into the tin shed. In a typical, functional tin shed, there are two major machines to be found – the separator and the air float. In some very big tin shed, there may even be crushers and the sieving machines inside the shed, while the jig machine is always located outside the shed.

In summary, the process of separation and floating by the separator and air float respectively are the major event in the removal of impurities without human intervention except in adjustment or setting of the machine. Both machines come in sizes and types. In most cases, washed or unwashed Tailing material are first floated and before running it on the separator.

Gravity Separation using Air:

This was carried out to separate minerals with different specific gravities such that the denser minerals were separated from the less dense minerals. The gravity separator machine used here made use of air as means of separation. It consists of a deck in form of a table with shallow corrugations, a moving belt which provided the mechanism that vibrated and provided the vibrations that shake the table and an air system created by a vacuum pump to separate the clustered particles. The material was fed into the machine through the feed. The table surface (deck) was inclined at an angle of about 10° to the horizontal with shallow corrugations running along its length at right angle to the direction by providing a rough surface that helped to group the minerals based on their density and specific gravity. The outlets consist of six openings with buckets lined up under each outlet. Mineral that dropped in buckets 1 and 2 in front and of the deck was high grade tin concentrate and did not require further separation. Buckets three to five contained low grade tin ore, high grade columbite, silica, titanium and illemite and other associate minerals hence require further separation. Because most of the associate minerals are magnetic in nature, a magnetic separator machine was used in further concentration in process. Bucket six was thrown out because it did not contain valuable mineral.

(A) **Air floating Machine:**

The floating process being the first in the machine processes usually is done using the Air float machine. As mentioned earlier, the Air float is an electromechanical machine which is made of solid metal body. The Air floats' configuration technical set up gives it an approximate height of 1.68meters and a width of 1.4meters, though not really square or rectangular in shape.

There are many technical parts which make up the Air float machine. There is the main platform where the Tailing material is floated. The platform is a kind of wooden mesh network which is covered with a yard clothes using staplers pins along the wooden grids. The wooden mesh allows air from the air bag to flow through it to the yard and subsequently to the material on top of it. The air bag itself (made with Trampoline material) is connected to the air duct section via the outside. The air duct is rectangular in shape and is a part of the machine body. It has a turning handle which adjusts the air inlet either by closing or opening a door or cover to the duct. The air float machine can also be inclined to any specific angle to allow for proper floating depending on the type of material and the expertise of the operator.

There is the speed adjustment part attached to a shaft which can reduce or increase the speed of the Air float machine and turning a handle will thereby simultaneously adjust the vibration of the Air float. The other part of the Air float is the electric motor without which the machine cannot move. The Air float machine has six exit points or funnels where already processed Tailing material dropped into pre-arranged buckets to be evacuated.

Due to the height of the Air floating machine, a strong wooden table–like platform is construction in an L-shape with many legs and a flat top to accommodate all the Buckets or Pans, it also served as the standing platform for the operators to work. The two funnels are located separately from the others at the head of the Air float machine and they are the most important of all. These are buckets 1 and 2 according to importance. There is also a place at the extreme left hand side of the machine (at the edge of the platform) where the fresh material to be processed is poured. This contains an opening which can be adjusted vertically upward to allow the materials to run through the yard platform.

(i) Big Air Floating Machine; There are two major types of Air float machines, the smaller types and the bigger types. Except for a keen observer, otherwise, one can barely understand the difference between the two types of Air floating machines. Some are locally fabricated or assembled while most are imported.

Majority of the imported Air Float machines are brought in disconnected parts and then assembled anywhere it is to be fixed and used.

The working principle of the Air floating machine is that of air velocity impact through movement in a controlled environment and that of vibration mechanism. The supplied electrical power allows the Air float to vibrate and with the vibration resulting in the separation of the constituent minerals that make up the ore. These minerals separates according to their specific gravity or body mass, and with the introduction of air into the system, it pushes the Tailing material forward through its velocity impact. The action encouraged the heavier ones to continue their forward acceleration while the lighter materials on the other hand are blown off the trail and flows downward as a result of the combined effect of vibration and air power. The various materials will then flow via the exit funnels and be collected at the different removable buckets or pans.

Since the buckets are numbered 1 to 6, buckets 1 and 2 are the higher grades while buckets 3 and 4 are the middle low grades and will either have to be refloated using free floating methods or they will be run in the Separator machine while some miners will prefer to wash it in Jirgi to improve its grade substantially before re-running or re-floating it. The last two buckets which are 5 and 6 will be thrown away as waste because it contains high quantity of impurities such as sand, stones, zircon, and monoxide and very low grades containing 65% - 75% iron.

Depending on the material, some high productive Tailing material when floated, the number 1 bucket will give a high grade Tin ore of say 17.7-18.0% quality grade which by all standards is regarded as high grade. The number 2 bucket will give 18.0 -18.4% quality grade which is the highest grade for a pure Columbite ore.

In other words, for a Very rich and highly productive Tailing material, the following will be true for Air float buckets arrangement when floated viz;

* Bucket No 1 = 17.4 – 17.9 Good grade Tin
* Bucket No 2 = 18.0 – 18.5 High grade Columbite
* Bucket No 3 = 18.9- 19.6 Middle grade Columbite
* Bucket No 4 = 19.7- 20.6 Low grade Columbite
* Bucket No 5 = 20.9 – 22.0 very low grade Columbite with high iron content
* Bucket No 6 = 22.1- 25.5 Iron, sand, zircon, and other waste impurities

(ii) **Small air floating machine:** the difference between this type of air floating machine to the bigger one is just size. This means that it can accommodate less quantity of material at any given time of floating, particularly free floating periods. Some of this type of machines are locally fabricated or assembled from parts which are directly imported from abroad. Majority of the imported Air Float machines are brought in disconnected parts and then assembled anywhere it is to be fixed and used. The small air floating machine share similar working principle with the bigger type as explain earlier. The major difference between the two types of air floating machine is that the small air floating machine has a smaller size electric motor with less horse power rating than the bigger air floating machine. The buckets are also numbered 1 to 6, just like we have in the bigger air floating machine. The air vent aperture is smaller when compared to that of the bigger air floating machine.

(iii) **Air Floating Techniques:**

There are two types of techniques adopted for air floating Tailing material in an Air float machine. These are the hand brush floating and the free floating techniques.

(a) **Free Floating technique:** The free floating method is started by the tin shed operator or in his absence by an expert who will set the process rolling. This type of floating method is carried out for large quantity of materials which could be wash or unwashed and may be as many as 4 to 15 trips. The Tailing material is first poured into the floating platform through an aperture and the operator will begging by initial floating technique until the material is set and working normally before he will leave it to be running on its own unaided for hours or even a whole day or days, until the whole material finishes. In this floating method, a person is dedicated to periodically and consistently feeding the Air float machine through the little aperture set for that purpose. The initial setting done by the operator for free floating may be dislodged out of setting once the person feeding the machine delayed for some time and the materials on the platform finishes. Another reason why a free floating material may be dislodged out of setting was when too much dust particles have blocked the tiny air spaces on the yard material and air cannot garner enough velocity to penetrate the yard material to create an effect.

In this respect it will have to be reset by the operator again and in the case of dust blockage, the operator will ask the person feeding the Air floating machine to stop feeding it. Then he will clean the yard covered platform and clear the dust away before the floating process starts again. One, two or three other workers will continuously serve the machine by evacuating the resulting buckets from 1 to 6 by placing them at their respective points either in marked bags or other positions inside the tin shed.

(b) **Hand Brush Floating technique:** the brush floating method is a regular floating method carried out on smaller quantity materials especially semi pure Tin ore, semi pure columbite ore, low grade columbite, washed monoxide, tantalite ore or wolfromite ore.

It is always executed by the tin shed operator, who uses a portable hand brush (made of reinforced polyvinyl chloride material) to control the floating process using maneuvers and techniques necessary for a successful processing. The operator may be required to use various techniques acquired through training and or experience over the years during practice. To conduct a successful operation, the operator may, depending on circumstances that prevails at the period of floating, adjust the air float machine platform to an incline angle of between 30 to 48 degrees or may open or close the main air vent by turning a screw handle or may increase the vibration of the Air float machine by adjusting some side buttons on the body of the Air float machine.

Here, the person feeding the Air float machine with tailing material will only pour at times when the tin shed operator or air floater told him to do so. Most at times, the quality of the finished product depends on the expertise of the air floater. That is why some air floaters are more popular than others and are coveted in every tin shed and they are being lured to other tin sheds by the managers through high remunerations packages.

Magnetic Separation Using EMF effect:

This was done to separate the minerals of high magnetic susceptibility from those of low magnetic susceptibility. A high intensity electromagnetic separator was used. Direct current (DC) was converted to magnetic field with the aid of magnet. The magnetic field created was then reacted with the magnetization of the ore that was fed through a process of lifting effect and pinning effect.

Minerals with the magnetic property were lifted by the magnet as they passed through the conveyor belt thereby falling into their respective buckets, while minerals with low affinity to magnetism remain pinned on the conveyor belt and dropped at the front end of the belt. The magnetic separation machine consists of different intensities increasing from the lowest intensity to the highest, a feed, a conveyor belt, control switch and control panel where the intensity of the magnets can be increased and an electric motor which provides the motion needed to move the conveyor belt.

50 kg of the feed was fed into the funnel shape part of the machine and into the conveyor belt which conveyed it through the various magnets. On this machine, which works by electromagnetic field effect, the material runs down through the conveyor belt and through the disc. At the first disc, magnetic and non magnetic minerals are

separated. The magnetic mineral, iron ore, which is the most magnetic is first separated on the first disc, columbite and other less magnetic mineral are attracted by the second and third discs. The non – magnetic mineral i.e. Cassiterite or tin ore (SnO_2), monozite, zircon sand in traces, zinc and other non – magnetic ore in the mineral goes to the front (i.e. they are not being picked by any of the disc). These are known as over – belt materials. Minerals such as hematite, limonite with high magnetic susceptibility was picked by the lowest intensity magnet, then followed by columbite, Cassiterite (tin) ore and then zircon which has very low magnetic susceptibility were collected in the front of the conveyor belt.

(B) **Magnetic Separator Machine:**

the separator machine does the actual separation of the constituent minerals. There are two main types of separators- the Rapid and the BMF. Rapid separators are smaller separators; they perform almost excellently in the removal of iron from materials. They have only the wide horizontal belts.

The BMF is a big separator that has both the wide horizontal belts alongside perpendicular belts. They too work well in the removal of iron and the separation of sand and monoxide from materials.

For both types of separators, they have similar (though not dimensionally equal), technical parts namely electric motor, magnetic belts, electric coils, magnetic discs, big rollers, setting knobs brush, long shaft, control panel, side funnels, cylindrical feed or silo etc.

The electric motor used depends on the size of the separator machine. The bigger the separator machine, the bigger the electric motor and its horse power rating capacity and vice versa. The magnetic belt is usually black in color as depicts most magnets in existence today. It is about 40 to 45cm wide and its wound round about the big circular revolving rollers which rotates in an anti-clockwise direction carrying the materials on it. The electric coils play a significant role in the separator machine. Current is normally supplied to the coils and the rotating magnetic coils from the power mains. This creates an electro-magnetic field which makes it possible for the different constituent particles in a material to be picked by the rotating discs. The rotating magnetic belt moves over or is placed on top of a flat plate metal which sits on the coils. The coils may be one, two or three under the rotating disc. The magnetic disc got its name probably because it is responsible picking out the mineral ore, although when seen closely enough, no visible magnet can be seen on its spiral teethes. The disc rotates with the aid of fan belts connected to the main rotating shaft. The rotating disc can be lowered down close to the materials or upward away from the moving materials on the belt. The two big rollers are situated at the extreme ends of the machines. The rollers move the magnetic belt around in anti- clockwise direction. The setting knobs are located on top of the separator with hard plastic round head to protect operators hand when turning. There is a long metal handle which is linked to the rotating discs. By turning the knob up or down means raising the disc vertically upward or downward. There is a small hook-like metal on the machine attached to the turning knob; this metal serves as an indicator to show the levels of drop or lift of the disc. The metal hook moves horizontally forward or backward along graduated markings in series showing the levels. A brush is located on the Separator machine at the front end, very close to the roller at that position. The brush is fixed close to the rolling magnetic belt to let the already processed material to be discharged inside the pan or any container that is placed at the front of the machine for that purpose. It is a wide brush covering the entire width of the belt, placed perpendicular to the direction of movement of the belt. The shaft of a separator machine is long enough to reach almost the whole length of the machine. It is a cylindrical metal with a diameter of about 6-7cm thick (this depends on the size of the machine).

For the Rapid separator machine, the shaft enables movement of other parts of the machine by transmission through belt drives. Another important component of a separator machine is the cylindrical bucket or open silo as it is sometimes called. This bucket is situated at the top of the machine, clearly the highest point of the machine. It is in the bucket or open silo that Tailing materials to be run are poured into. The top is open without cover while the lower end of the bucket or silo has steep edges narrowing to culminate into a small opening so that the Tailing material poured into it can roll down to the end itself without any need for interference. This opening can be controlled by reducing or increasing the tiny opening to allow the materials to drop onto an inclined flat steel plate of a width equal to the size of the magnetic belt. This flat steel plate enables the Tailing materials dropping from the controlled tiny opening to be spread wide enough to fit the width of the rotating magnetic belt.

The control panel of the Separator machine is fixed on a wall close enough to the Machine itself. It plays two key significant roles; firstly, it enables operators to view the analogue meter readings to know the amount of electrical current available and also employed by the machine coils at every given time. Secondly, it assist the operator to increase or decrease the current to suit the type of setting needed on how to separate certain type of materials ore. Beside every meter reading display on the Separator control panel is the turning knob positioned at the bottom center of the meter, close enough to see the reading and adjust it by turning the knob. The control panel is used to increase or decrease the current flow to the coils.

It will be noticed that by increasing the electrical current of the coils, the magnetic influence or flux is increased and the machine will pick a lot of iron to the side including some good minerals. Also when the electrical current is reduced from the control panel, the magnetic effect will reduce. Setting an optimum and suitable amount of current to the coils for dressing different mineral ores will take an experienced operator to control. The operating principle of the Separator machine is on electromagnetism, though the embodied machine is an electromechanical as well as electromagnetic machine which takes advantage of the natural magnetic affinity in the properties of most naturally occurring metals. This property of the metals is utilized in the separating process. When the electric current passed through the coils, it induced magnetic field which has a region of flux that enables the mineral ores to be fused and separated (by the aid of the rotating disc).

The main types of magnetic separators are the BMF, the Rapid and the Single Disc Separators.

(i) **BMF Magnetic Separator;** this type of separator machine is very big, some are six meters(6m) long, and two meters (2m) wide. In addition to the horizontally moving conveyor belts rotating on two extreme end giant size rollers, the BMF Separator also have about three other vertically rotating fan belts or chain drives covered.

This three vertically rotating drives are each meant to supply mechanism for the three very big and heavy rotating discs. It has exit funnels in opposite direction at the side of the separator machine equal to the number of coils present.

(ii) **Rapid Magnetic Separator;** this type of magnetic separator machine is the most common, it principally contains the main magnetic conveyor belt revolving as a result of rotating rollers at extreme ends of the machine. Some of the Rapid magnetic separators have three discs while others have two discs. It also has exit funnels at opposite sides of the machine depending on the number of discs on the machine. Some Rapids magnetic Separator machine has vertical transmission gears just like for the bigger magnetic separators, but are fewer and smaller when compared to the big Separator machine.

(iii) **Single Disc Magnetic Separator;** this type of magnetic separator machine is an eye sore. It looks attractive to the eye. It occupies little area space, just about 1.85m by 2.0m. This type of Separator machine has a single, but giant size disc which is very heavy and occupies most of the central working space of the Separator. It has a wide magnetic belt that rotates on a single large roller (or in some of this type of machine on double rollers) assisted by vertical belt drive at the side of the machine. The machine has one exit funnel at opposite sides of the machine.

Types of Separator Machine Dressing; There are two major types of dressing or processing Tailing materials on a Separator machine.

(i) **Passing:**

This process is carried out on mostly large quantities of raw Tailing materials. This is done after a dug or washed material is perceived to contain large quantities of impurities in it, particularly the dreaded iron ore. During the process of passing a Tailing material, all the Magnetic Separator discs are dropped lower, closer to the revolving magnetic belt. This will mean that the forward movement of the Tailing material is done to remove iron ore from the material at all the side pockets, here only iron is removed while sand, Columbite, Tin, Monoxide and Zircon are transported on the magnetic belt and deposited at the pan in front of the machine.

The Separator machine discs will be set to pick only iron ores from the Tailing material and this process is always set to be very fast because the small opening from the bucket or silo will be open wider to allow the passage of many materials to the revolving belt. The experienced operator will reduce the current flow to the magnetic discs on the control panel so as to avoid or reduce the picking away of all the magnetic tin which are always disguise in iron ores. When iron is removed in such manner, sometimes little amount (about 2%) of columbite and monoxide are picked and discharged along with the iron ore. As mentioned earlier, some irons removed through this way are actually not iron but tin ore referred to as "magnetic tin". Sometimes, after the running, all the iron will be packed and floated to remove all the magnetic tin.

(ii) **Running:**

The types of material that qualifies for running are those that have already undergone the initial floating process whether free floated or hand brush floated. During running process on a magnetic separator machine, the first disc can be set to remove the remaining iron ore in it, because majority of the iron ore has been removed during "passing" process.

The second and third discs are then lowered down to remove columbite. While this is being done, the remainders mineral transported to the front belt to be discharged for further processing are sand, zircon, tin and some monoxide while some monoxides are picked along with columbite to the side pans during the passing process. The sand, zircon, tin, and monoxide collected at the front belt will be taken to the Air float machine to be floated so as to separate pure tin ore from sand and monoxide. This will be done by the tin shed operator using hand brush.

(iii) **Re-Run:**

During Re-run process, all the columbite ore that was picked by the magnetic discs and collected to the side pans during the "running" process has some amounts of monoxide in it and will have to be re-run again in the a magnetic separator machine to totally eradicate the monoxide and separate it from the columbite. This process entails an experienced tin shed operator to control the following procedures:-

*material flow speed from the silo or bucket,

* input current to the coils from Control Panel, and

* positioning of the Discs relative to the magnetic belt.

The technique of Re-run is to drop the first Disc slightly lower to the revolving magnetic belt and to reduce the current to the coils for that disc so that it can pick some hidden and stubborn iron ore still remaining with the Columbite. The purpose for reducing the electrical current flow to the coils in that position is to prevent the magnetic disc from picking the columbite ores to the side pan.

The second and third discs are lowered well down close to the revolving magnetic belt, but with the electrical current slightly increased to the coil. This is meant to draw and pick all the Columbite to the side pans while the remaining impurities containing 78% monoxide, 13% Zircon and 9% sand are left to continue its forward movement to be discharged at the front pan. This is called "front belt material". This same front belt material will be taken back and poured into the silo the second time to be run again to further remove the columbite still present in the front belt particularly attached closely to the monoxide.

This last process of re-run will produce very good and high quality grade columbite with shiny sparkles which is also very heavy as compared to the earlier separated columbite. The relative heavier weight of this Columbite is due to its closeness to monoxide, because a typical monoxide is heavy.

Whether dressing a material using the Air floating

machine or dressing it using a Separator machine, none is independent of the other for a successful completion of the process to produce a final product. At the glory days of mining activities, during the peak of tin and columbite processing on the Plateau, some customers will have to wait days on end, even a week to be able to dress their Tailing materials. This is because there are so many miners coming from far and near queuing and waiting for allocated dressing time. In this wise, any of the machines can be used first depending on which is available at the time. Sometimes, even the tin shed operators are allocated time to dress their "swept base materials" (This depends on the disposition and leniency of the manager and their relationship to him). This so called "swept base materials" are very rich because it is a combination of different materials that fell beneath the

Separator machine or got stuck inside the Air float machines during separating and air floating.

POST-PRODUCTION STAGE:

[3] Post-Machine Processes:

This stage could also be referred to as the quality assurance stage. All the activities here are carried out to ascertain the true value of the resultant finished product and ensure its marketability. Post machine processes are the activities carried out after the material has been refined through the initial processes of pre and machine processes. Here the authenticity, genuiness and quality is checked to determine its price. The process is mostly carried out inside the chief operator's office or the manager's office.

This stage is so crucial that any finished product that fail to give the desired quality could be taken back to the machine for a repeat dressing and to still undergo further dressing processes.

(a) **Mixing and Taking Samples:**

(i) **Hand mixing;** the main reason for mixing final products is to pick samples and test it for quality assurance. Hand mixing is mostly carried out on materials that are few in quantity. Mixing final product needs someone with experience. The mixer uses both hands to mix the product and a combine mixers of 1 to 4 people can mix together at the same time on a single piled material depending on the urgency and need of the moment. The final Tin or Columbite ore to be mixed will be placed on a well swept, smooth floor. Bare hand or hand glove could be used to mix the mineral ore together. Two to four bags of material could be mixed together with hand successfully, but any quantity above four bags will not be suitable for the hand mixing rather it will prefer the shovel mix. After a thorough hand mixing, of tin or columbite finished product, samples will then be picked from three, four or five different positions from the mixed product and then taken to be burnt in case of Tin or to be burreted in the case of Columbite.

(ii) **Shovel mixing;** as the name implies, this type of mixing requires a shovel to be used due to the large quantity of finished product involved. One, two or more people can do the mixing here together, just like the hand mixing. But for the shovel type of mixing, it requires large quantity of finished product. Most at times, this mixing is done to properly blend together concocted finished products together, example when a burnt low grade columbite is mixed with another high grade columbite, the mix will have to be done thoroughly enough.

(b) **Acid and Powder Burning** (for Tin alone);

(i) **Chemical Acid**; The chemical acid is used alongside the powder to burn the final Tin ore which is already processed. This type of acid is concentrated Hydrochloric (HCl) acid found in most laboratories. During the test process, the powder is first mixed with the tin ore before the gentle application of the acid and stirring gently for some time.

(ii) **The Powder**; the chemical powder is an ash colored dry powder substance which comes in a small, round plastic container. This powder is used to test the final tin ore.

(iii) **The Burning Process;**

A sampled Tin ore will be placed inside a flat or nearly

flat plastic plate and the chemical powder is poured on the tin ore sample. The forefinger is used to mix the two materials together thoroughly until the whole tin ore sample inside the plate takes the color of the powder. After that, an acid is poured on the mixture. The plate is then vibrated by shaking it with one hand from side to side to ensure that the acid mixes well with the other content of the plate. The person carrying out the burning dare not use his finger to do the mixing because of the acid's concentration. It will be observed that immediately the liquid acid was poured on the mixture, a chemical reaction will ensue with conspicuous boiling which exhibits a visible white gas and a pungent odor that irritates the nose and even eyes if brought closely to it. When the obvious reaction stops, the Tin ore is washed with clean water using the forefinger until all the burnt chemical powder is washed away.

After the washing, when viewed closely, it will be observed to have burnt into a whitish Aluminum color. When all the burnt tin ore inside the plate displays all round whitish substance, then the ore is indeed tin and it will sell. Otherwise, if there is a mixture of burnt ore with Aluminum color and some black color combined together, then it means the tin ore is not 100% tin and a further process is needed to separate the foreign elements. The percentage of the black particles will determine the type of re-dressing to be carried out on the ore. Most at times it will have to be "passed" through the separator machine as explained earlier to remove the black particles which could be Columbite or iron.

If on the other hand the tin ore sample refuse totally to burn into Aluminum color substance, then it invariably means that the sample is not tin ore at all or rather either of the burning chemicals is defective. The resultant burnt ore can be rated in percentage based on the quantity of black un-burnt substance in the sample. Sometimes, when the un-burnt particles are just about 3-5% of the whole, the buyer may use his digression to buy at certain discounted price and later mix it with other larger quantity of clean tin ores which will make the black particles to disappear in the gross mixture. Sometimes, Columbite is also burnt to check for residual Tin content in it. Though not necessary, but is being done as a check for any double standard activities by some miners through grade boosting by using Tin which is a relatively heavier mineral to boost the grade of columbite by adding to its weight. The test-burning for Columbite is similar to that of Tin using Acid and Powder. If after burning Columbite and it is discovered that the whitish Aluminum content in the burnt sample is around 30-35% of the sampled material, the Columbite ore will be referred back to separator machine to be removed or extracted through the earlier explained process of running.

During running in this case, the Tin ore can be separated from the Columbite ore by dropping the Disc hand of the Separator machine lower. This will make all the columbite to be picked and drawn to the side pans, while Tin will run on the conveyor belt and be collected at the front pan. Most at times, when Tin ore is deliberately added to Columbite to boost its grade, and after the Tin ore is separated from the Columbite, it will be observed that the grade of the Columbite ore will drastically fall down by a wide margin even as wide as 5 to 10 points lower from high to intermittent or to low grade.

Example,

a dressed Columbite ore sample, which gives 19.2 grades after Burreting, but fell the Acid and Powder burning Test (with evident display of Tin ore) is referred back to the Separator machine by the Manager in charge to be re-processed by the Operator on duty. After the Separation was done, the Columbite is mixed and a sample is taken for Burreting. After Burreting, the grade now becomes 20.0, which is an 8 point down grade when counted backwards from 19.3, 19.4, 19.5, 19.6, 19.7, 19.8, 19.9, and 20.0.

It was therefore concluded that the initial Columbite ore was deliberately concocted.

(C) **Burreting or Volumetric Analysis** (for Columbite and sometimes Tin):

For columbite, the method of testing the finally processed ore is referred to as "Burreting", such as is done in most chemistry laboratories in Senior secondary schools, but in a different way. Those laboratory equipments which constitute the main backbone for the Burreting of Columbite are the Scale Balance, the Burreting tube, the Tub holder stand, the Plastic cup, Funnel and the Cleaner.

(i) **The Burreting Tube**; The Burreting tube is a slender cylindrical transparent reinforced glass tube of approximately 20-25mm in diameter. The glass tube has clear graduation marks at one side from top to bottom. The cleaner is a long slime metal with soft foam around its circumference which can go down into the tube and dry clean water molecules inside the tube. The glass Burreting tube has to be thoroughly dried inside, except for the level of liquid volume inside. This must be so, to prevent the mineral ore from being soaked, absorbed and stuck to the side wall of the tube, thereby reducing the actual quantity of the balanced and weighted Columbite ore necessary for successful Burreting. A strong rubber is attached to the end of the Burreting tube and clipped at the tail to prevent the liquid inside the tube from escaping. At the end of every Burreting process, the clip holding the rubber is removed. This allows the admixture of both liquid and the mineral ore inside the slim hollow tube to be dislodged away to a readily positioned conical flask bellow it.

(ii) **The Tube Holder Stand;** this Tripod Stand is a vertical steel unit that has a wide stable base and a horizontal gripping arm. At the end of the horizontal steel arm is an adjustable hook which has a small steel butterfly control turning knob at the side to open or close it. This hook help to hold the Burreting glass tube vertically in place. Sometimes this hook will be open to be able to remove the Burreting tube for maintenance or replacement.

(iii) **Plastic Cup;** this is a very small plastic cup which is approximately 50mm-80mm in diameter. This small and sometimes overlooked container is very indispensable because it has to be a material that is almost weightless or otherwise a material whose weight is insignificant. Steel or other materials are not suitable for this purpose because of rusting or extra weight. The Plastic cup will have to be initially weighed for default balance from inception or whenever a new cup is introduced as a result of lost or damage.

(iv) **The Scale Balance;** There are varieties of spring balances used in the mining industry for the Burreting of columbite and other mineral ores, these amongst others are as follows

(a) **Obsolete Justice Scale Balance:** This type of scale balance has reigned and was widely used at the early period of mining on the Plateau. The scale balance has since disappeared in the mining scene and became a scarce commodity. The earlier ones used are those that have two arms like the justice symbol. This type of Scale balance, has one of the arms holding a measured weight or a standard weight while the other arm contain the trough where the material to be balanced is poured in to. Each of the arms has a hanging rope or chain holding the trough at one side (left side) and the weight standard (right side) at the opposite side.

Processed mineral ore will be poured gently and gradually into the trough using a spoonful being held by the left side arm of the justice-like scale, until when the two arms settles at the same level and stops swindling up and down, in this case the mineral ore sample is said to be balanced. This type of spring balance is placed inside a transparent glass box to prevent wind action (it is sensitive to air movement or atmospheric pressure) from distorting the spring balance result during balancing procedure.

(b)**Modern Spring Balance:** this is delicate spring balance equipment that is sensitive to even a single grain of mineral ore. The modern spring balance has a height of 0.18m – 0.2m and a length of 0.45m. The left side of the Spring Balance has a circular flat top base with a diameter of 0.17m. This circular flat surface serve as the base for placing the Plastic cup or container which contains or carry the mineral ore, it also doubles as the position of the Spring which is of course situated hidden under the top flat circular base. This gives it its sensitivity to specific gravity and weight of minerals. At the right side of the spring Balance is the position for attaching independent predetermine weight blocks which comes in small rectangular blocks with one face formed as a hook for attachment to the machine. Some of the common weights are 50grms, 80grms, 100grms, 120grms, 150grms, and 200grms and so on.

These weights are used to initially set the default Spring Balance standard prior to introduction of any external weights on it. An extension in the form of a long arm from the left side of the spring balance extended to the right side, the end of this long arm has a middle dashed mark which is a pointer to indicate a complete balance of the machine either during the initial default standard balance or during the mineral ore balance.

The long arm, being attached to the spring section gently moves vertically up and down in both case of balancing process as mentioned above.

(v) **The Burreting Process:** at the peak of Tin and Columbite mining on the Plateau, Tin is usually tested through acid and powder burning only while Columbite is Buretted. But recently, some mining companies and or Directors sometimes insist on Tin to be Buretted just like Columbite ores. This may not be unconnected to modalities relating to exportation.

Equipments Setting: The first step in Burreting is to set the equipments ready in place. The clip holding the rubber at the end of the Burreting tube should be checked and tightened to make sure it is firmly clipped to prevent liquid movement and escape in that direction. Next is the pouring of liquid substance into the glass Burreting tube. The most widely used liquid substance is the "Premium motor spirit" popularly referred to as "Petrol. Petrol is commonly used for Burreting because of its inherent characteristics which are suitable for this purpose. A specialize equipment is used to contain the Petrol in it. This container is a plastic bottle like a Baby feeding bottle, but in this case, it has a long thin narrow mouth just 4-5mm in diameter and this long narrow mouth is bent over like L. Petrol, from the Plastic bottle is poured into the Burreting glass tube to a specific marked level and the initial reading is noted. After pouring Petrol into the Burreting tube, the long narrow cleaning wire which is surrounded with foam or yard material is used to clean the areas which are suspected to be stained by Petrol during the downward pouring.

Mineral Ore (sample) Setting: A sample of the mineral ore to be Buretted is well mixed and fetched in a small plate, it is from this plate that a small steel or plastic spoon is used to fetch some little quantity of the mineral ore and pour it inside the plastic cup. When poured, the weight will create an effect on the spring arm which will bounce up and down continuously. Intermittently, some little amount of the sampled mineral from the plate will continue to be poured onto the plastic cup gently until the spring arm stops moving up and down and balances with the mark on the right side of the spring balance machine. Once this is done, the mineral ore is ready for Burreting.

The Burreting Sequence: The Burreting sequence referred to an activity that proceeds equipment setting. It is therefore the actual Burreting process itself. The process is short and brief once the previous steps are carried out. After taking note of the initial reading, the mineral ore from the plastic cup which is set out on the Scale balance is now poured into the Burreting tube through a small funnel which is placed on top of the open top of the Burreting glass tube.

Immediately this is done, it will be noticed how the mineral ore will rush down the tube and will display setting quantity of Petrol in the tube. In a swift reaction, the momentum of the falling mineral ore will cause a stir in the volume of liquid inside the tube; this will be allowed to settle down before a clear reading will be carried out. The person carrying out the Burreting should apply a little tapping to the side of the Burreting tube to allow quick dissolution of the dry sampled mineral ore inside the liquid substance. The dropping of the particles into the tube containing Petroleum fuel displaces the fuel upward, depending on the weight of the material. For instance a heavy material will just settle quickly at the bottom of the tube and displace little quantity of fuel upward, while on the other hand, a light material or low grade will continue to rise up due to its weightlessness in fuel and will displace more fuel upward.

In rare cases, water is used as the Burreting substance especially in the absence of Premium motor spirit or Petrol, but the authenticity of the result is questionable. For example, a balanced quantity of Columbite is to be Buretted. If the fuel's initial reading is set at 12.0 marks on the graduated Burreting tube, and if after the Burreting sequence the liquid fuel is displayed up to the 30.0 mark on the graduated Burreting tube, the final reading will be taken after the reaction stops. Here, the top level of the Petrol mark will become the new and final reading.

Then the final grade of the Columbite will be calculated using simple mathematic as follows

Grade = Final Reading − Initial Reading

30.0 − 12.0 =   18.0

Any Columbite with such reading is considered to be a High Grade Columbite.

(D) Analysis (Tantalite, Columbite, Tin etc)

(i) **Introduction;**

in mining, analysis is carried out to expose the totality of a mineral ore composition through x-ray. ED-XRF Chemical Composition of Cassiterite Deposit X-raying of mineral ore samples only came about in latter stage of Tin and Columbite mining on the Plateau, particularly at the onset and peak of Tantalite mining. X-Ray Fluorescence (XRF) is a technique with the ability to deliver fast and accurate results with little or no sample preparation in various stages of mining activity from grass root exploration to exploitation, ore grade control, and even environmental investigations. Portable x-ray fluorescence (XRF) instruments are great tools not only for prospecting, but also for grade control of these metals. Accuracy is a key factor in the evaluation of ore concentrates, particularly due to their high trading values. When a finally processed columbite or Tin ore is presented, it will seem as if it was in its purest state without any conspicuous mineral element in it, but after analysis was carried out, most miners are perplexed to discover that what they thought to be pure columbite or Tin turns out to contain several other mineral ores in various quantities.

For example;

a well dressed or well processed Columbite sample given as 18.1 (high grade Columbite) after Burreting will present the following result after analysis

Columbite (Nb, 75.4%)

Tin (Sn, 1.56%)

Monoxide (Mz, 3.2%)

Zircon (Zr, 6%)

Iron (Fe, 3%)

Wolfromite (W, 3%)

Tantalite (Ta, 2.74%)

Lead (Pb, 0.37)

Majority of miners prefer the modern Analysis method

rather than the normal traditional Burreting method. In the example given above, a miner who depends on Burreting of his Columbite will end up being paid for the high grade columbite only, while another miner who prefer to carry his material to the tin shed that has and uses the analysis machine will receive payment for the useful minerals ores in the sample according to their corresponding percentages as displayed on the output receipt from the analysis machine print out. From the example given above, the miner will be paid for columbite at 75.4% of the total weighted material, Tin at 1.56% of the total weighted material, Wolfromite at 3% of the total weighted material, and Tantalite at 2.74% of the total weighted material. The others which comprises Monoxide 3.2%, Zircon 6%, Iron 3% and Sand 5% are unwanted impurities and none-valuable mineral ores which will not receive any payment except in some rare cases when they are required for some discovered specific usage.

(ii) **Analysis Equipments/Machines:**

(a) **Portable Handheld XRF Analyzers**; the following are some of the most common portable hand held analysis machines in the mining industry namely:-

1. Niton XL2 GOLDD

2. Niton XL3t GOLDD+ Series analyzers

3. Niton FXL field-mobile x-ray lab

these machines deliver fast, accurate elemental analysis with unmatched efficiency across all stages of mining and exploration. They combine geometrically optimized large area drift detector (GOLDD™) technology and the high-power x-ray tube.

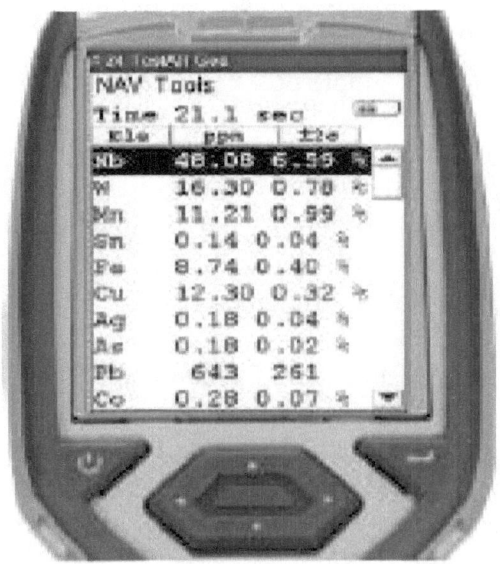

The Thermo Scientific Niton XL3t GOLDD+ XRF analyzer delivers fast analysis of Coltan. Minerals Samples analyzed with a Niton XL3t GOLDD+ portable XRF analyzer, as well as those analyzed using laboratory methods through (ICP and AAS) for comparison purposes are discovered to give similar results. This is a fundamental parameter model that is able to quantify more than 30 elements, typically without the requirement of user calibrations. However, when required, a simple post calibration adjustment can be made to improve it.

Many minerals occur as dark gray to black grains which may look like Coltan in fine grains. The high correlation between assay results from the handheld Niton XL3t GOLDD+ and lab data indicates that it is possible to successfully identify high- and low grade concentrates of Coltan samples in seconds using portable XRF. Whether you choose the Niton FXL, Niton XL3t GOLDD+, or Niton XL2 GOLDD XRF analyzer, all Thermo Scientific portable XRF analyzers deliver fast and accurate elemental analysis for intensive metal exploration and production, from base metals to precious metals and even rare earth elements (REE).

(b) Bench top bulk analysis XRF:

Bench top Bulk analysis machines are bigger than the

portable hand held XRF analyzers. When closed and not in use, they resemble a hand held business briefcase. This category of analysis equipments are sizable enough to be placed on human laps or on table flat tops. Analysis with the highly flexible and powerful energy-dispersive X-ray fluorescence (EDXRF) spectrometer ranges gives supreme output and highly appreciative results. This XRF tools also ensure quality assurance and process control requirements across a diverse range of applications; these includes the following:-

(I) X-Supreme 8000 XRF Analyzer,

(II) LAB-X5000 XRF Analyzer, and

(III) LAB-X3500 XRF Analyzer.

(I) X-Supreme 8000 XRF Analyzer:

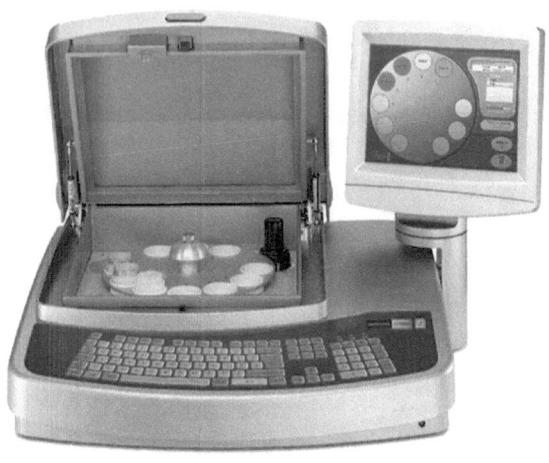

(a) The XRF analysis with the X-Supreme 8000 analyzer can be performed on a wide range of solids, liquids, powders, pastes, films, etc

(b) it requires minimal or no sample preparation

(c) the operation can be unattended and it can be used by non-laboratory operators.

(d) It has FocusSD technology for speed, accuracy and long-term reliability

(e) it has flexibility to perform qualitative, semi-quantitative and full quantitative analysis.

(II) LAB-X5000 XRF Analyzer:

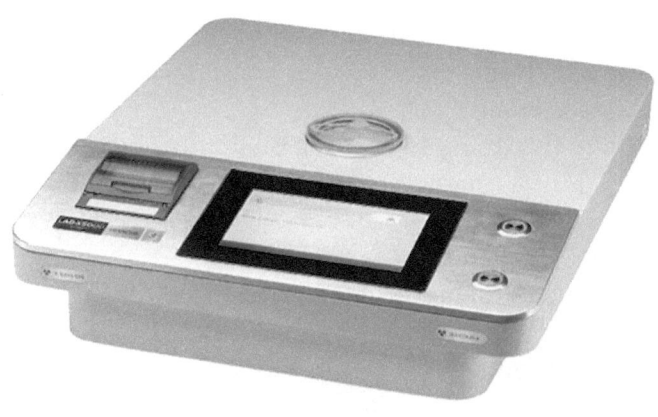

(a) they are Easy to operate,

(b) they need minimal or no sample preparation,

(c) they have low limits of detection,

(d) they have rugged and robust design for harsh environments,

(e) they have flexible software for user customization and method development,

(f) they have easily programmed instrument parameters,

(g) they have comprehensive qualitative or full quantitative analysis,

(h) they have flexibility to perform qualitative or full quantitative analysis.

(i) their software and hardware are optimized for high throughput testing,

(j) they have user interface which is inspired by leading line of point-and –shoot handheld analyzers,

(k) their large industrial-grade touch screen is resistant to chemicals and clearly displays results and controls, atmospheric compensation allows all measurements to be taken in air path, eliminating the cost of external gases.

(III) LAB-X3500 XRF Analyzer:

XRF analysis with the LAB-X3500 has the following defined advantages

(a) it saves time and money and increases productivity

(b) it is designed to perform in a wide range of locations wherever quality control is required e.g. in the laboratory or on site.

(c) It is a reliable and affordable XRF analysis for a wide range of industries such as petroleum, wood treatment, cement, minerals, mining and plastics.

(d) This model is optimized for analysis in air path thereby avoiding the requirement for any external gas.

(e) it is small and compact providing 24/7 peace of mind and ease of use by production staff.

(iii) Analysis Process and Presentation:

The X-ray fluorescence spectrometer (XRFS) was switched on and allowed to warm up and also gained in order to stabilize the optics and the x-ray tube. It was then calibrated to determine the expected chemical composition in the ore.

The analysis process is carried out for two main purposes namely

(a) **Offsite Analysis**; the offsite analysis is mainly carried out in the tin shed or in some external Laboratory with the availability of analysis equipment. The offsite analysis is carried out on a sample of a well processed Columbite or Tin ore to ascertain its value for the purpose of marketability. After all, the major objective of mining of Tin and Columbite is to make profit and have good return on investment. Tin and Columbite mining is one of the few business investments in Nigeria and even in sub-Saharan Africa that is capable of giving back ten(10) or more times return on investment. This business is capable of lifting a person's status from average to super wealthy overnight. This has been seen, witnessed and tested on several occasions. In this process, the sampled material needs to be well dressed or finally processed, if it is expected to attain its peak value for better income. A teaspoonful of the sample is poured into a small container which comes with the equipment. This in turn is mounted on the XRF Analyzer through a circular depression set for it. The sample doesn't need any initial preparation (that is, it doesn't need to be prepared into paste, solid or film) for the purpose of testing. The next step is to power the analysis equipment by pressing down a button at one side of the equipment then allow the machine to boot, recognize the presence of the sample through sensors, and then scan the sample through x-raying. Finally, the reading is displayed on the monitor; beside it is a question whether to print the result. A button is set for the printing of the displayed result and pressing the button will issue out the printed hardcopy in a little rectangular paper.

(b) **On- site Analysis;** as the name implies, this type of analysis is only conducted on a specific site. In this case, the miners travel with their portable handheld or desktop XRF analyzers to the source site, whether it is located at the excavation point or it is located at the processing zone outskirt. This is mostly situated at sites far away from machines and any kind of processing equipment. This type of analysis is carried out for the purpose of initial investigations, material sampling and research purpose.

The on-site analysis is performed on some samples of large deposits of raw Tailing materials, freshly dug mineral ores and surface and ground Tailing dumps to verify its content and study its viability for the purpose of investing on the material, in other words, the on-site analysis serves as an advanced pheasibility studies. In this process, the material to be sampled will first be picked at various points of the Tailing material (pile, dump, mound or heap) through a process of sampling referred to as "random sampling". The next step is to screen the samples through a hand sieve to remove impurities such as coarse aggregates, stones, little plants, organic matter, bottles, plastics, polythene material etc to prepare it for analysis. After the sieving, the material sample is properly mixed before the final sample is taken for analysis. The proceeding final analysis process using the different XRF Analyzers is carried out as explained above for the offsite analysis.

Example 1:

An on-site energy dispersive x-ray fluorescence (ED-XRF) analysis to determine the chemical composition of a tailing deposit from dogo na-hawa in Bukuru, Jos

South LGA, gives the following chemical compounds and its associated percentage (%) composition viz $CaO=2.8\%$, $TiO_2=16.7\%$, $V_2O_5=0.2\%$, $Cr_2O_3=0.93\%$, $MnO=1.1\%$, $Fe_2O_3=23.7\%$, $NiO=0.26\%$, $CuO=0.50\%$, $As_2O_3=0.01\%$, $Y_2O_3=0.52\%$, $ZrO_2=29.1\%$, $Nb_2O_5=8.9\%$, $Rh_2O_3=4.0\%$, $SnO_2=7.2\%$, $CeO_2=0.5\%$, $Yb_2O_3=0.46\%$, $HfO_2=0.81\%$, $PtO_2=0.31\%$, $PbO=0.33\%$, $ThO_2=1.6\%$, .

The above analysis result of the raw tailing material obviously indicated to any prospective investor that the percentage composition of Cassiterite ($SnO_2=7.2\%$) and columbite ($Nb_2O_5=8.9\%$) ores which are the main area of interest, exists in small percentage which in turn will determine the value of the tailing material and therefore will guide the investor on how to wisely buy the tailing material and process it to get profit.

Example 2:

A spectrometer and energy dispersive x-ray fluorescence (ED-XRF) analyses of a pit-dumped tailing material from Tsohon Foron in Barkin Ladi LGA, gives 40.4 % Nb; 24.8 % Fe; 21.5 % Ti; 2.3% Ta; 5.8 % Sn; and 4.9 % W the above stated analysis result indicates that the tailing material looks promising for investment not only for the two major areas of interest which is tin (5.8 % Sn) and niobium (40.4 % Nb), but also has advantage of the presence of extra mineral ores in it such as tantalum (2.3% Ta) and Wolfromite(4.9 % W) which translate to good prospect and sources of income.

(E) **Other Post-Production Tests** (Wolfromite & Lead)

(i) **Wolfromite Test;** one of the most important post machine test usually conducted on processed high grade mineral ores, particular columbite is the Wolfromite test.

Wolfromite is a relatively heavy mineral ore when compared to tin, columbite or monoxide. It has higher specific gravity than the earlier mentioned mineral ores. When Wolfromite is Buretted, it will fall into the class range of 15 to 17 grades. A high grade Wolfromite will give a Buretted result of 15.8-16.3 grades. Superficially, this mineral ore looks like columbite, but it has finer grains, greater sparkle and shinier luster than any existing columbite. At the peak of tin and columbite mining on the Jos Plateau, Wolfromite global economic value comes far less than that of tin or columbite, but its ability to boost the grades of low grade columbite to high grades makes it valuable amongst miners.

A closer look will indicate that Wolfromite is not totally black or dark blue in color, it has some little content of brownish particles in it and some shiny scales like that of fish body that stuck to a person's hand. When wolfromite is blended with columbite it boosts the columbite weight almost three (3) times and the eye can hardly tell. So to check for Wolfromite, the processed high grade columbite is subjected to a Wolfromite test. To start the test, the high grade columbite is well mixed and sample picked. A tea spoonful of the sample is fetched and poured into a test-tube which is less than 3inches. Some amount of liquid concentrated HCL acid is gently poured into the test tube containing the high grade columbite and mixed together. After that was done, a thermal pressure will be applied to the bottom of the test tube in the form of heat whether from a lighter, candle or a Bunsen burner. After some minutes of heat application, the mixture will begin to boil, after boiling; a close observation will reveal the true identity of the Columbite in that if the color changes from black or dark blue to light green, then it will be concluded that

Wolfromite was added to the Columbite. But if after the test, the original color of the columbite remains unchanged, then the columbite ore is truly a high grade one.

(ii) **Lead Test;** Lead is also a heavy solid mineral. Lead falls into the Buretted class range between 16 to 17 grades. It is totally unwanted in the production of tin and columbite. Lead in its natural form is a coarse black mineral with tetrahedral shape or pencil like texture. Lead is crushed in steel mortars or using the crushing machine to make it finer and appropriate for mixing with other mineral ores. Due to its denser nature and high specific gravity it is used by some miners to boosts the grade of tin or columbite. Tin, because it is none shiny and heavier than tin and therefore can blend well with tin or columbite and remain undetected until a test is carried out to manifest its presence if at all it is present.

The actual test for Lead is done by using a finally processed, well mixed and sampled high grade tin or columbite ore. The sample is fetched using a teaspoonful and then poured into a flat plastic plate. The reason for using a plastic plate is to prevent any chemical reaction that may occur between the material molecules of steel with that of the mineral ore and the chemical acid and powder.

The test process itself begins with mixing the high grade mineral ore inside the plastic plate with a chemical powder. The two are mixed thoroughly with the index finger before pouring the liquid concentrated acid into the mixture. Immediately an ensuing chemical reaction commences and burning took place. If Lead was present in the mineral ore, it will give out or release a pungent repulsive odor like that of rotten eggs which is an indication of the presence of Lead in the mineral ore, otherwise there will be no such odor and therefore no Lead.

(F) **Weighing:**

(i) **The weighing Scale;** the most common weighing scale used for weighing processed tin and Columbite ores is the circular, transparent glass front type with a hook. The Weighing Scale has standard factory graduations with measuring units in pounds weight and Kilograms. This type of weighing scale is the hanging type via a hook. The hook is positioned at the top side of the Scale for hanging. The hook is fabricated monolithically to the body of the weighing Scale. This weighing scale is hanged on the horizontal member of a fabricated steel post.

(ii) **The Steel Post;** the steel post has two vertical legs-stands whose measurement could vary depending on the intended functions. Most of the Posts in existence are 1.2m high by 2m wide. The steel post is made from hollow solid cast iron of 4-5mm thickness. It depicts a soccer goal post, a hang man's post or a butcher's hang which comprises two equal vertical poles with base legs and a single horizontal bar. It is the horizontal bar that accommodates the weighing scale, which is hanged or tied in position. The two vertical poles provide structural support and stability in transferring the weight to the ground through gravity.

(iii) **The weighing Bucket;** A standard fabricated steel weighing bucket is used for weighing both processed Tin and Columbite. The bucket is specially constructed to attain a defined self weight. The bucket has to be balanced on the weighing scale. The weighing scale pointer arrow is always initially below the zero mark on the graduated scale. When balancing the weighing bucket, the self weight of the bucket will raise the pointer arrow hand to the zero mark, which is directly vertical at ninety degrees (90^0)

(iv) **Standard techniques for weighing Tin and Columbite;** the standard weighing units for processed Tin and Columbite ore is the pounds weight (lbs). During weighing, the final processed Tin or columbite ore is poured inside the weighing bucket and together, raised to the hook of the weighing scale through the bucket's handle. The hook of the weighing scale will hold the bucket handle and invariably the bucket itself and its whole content.

(a) **Handle and Hook Technique** - here, the operative doing the weighing will normally grab the weighing bucket handle with one hand (probably a right hand) while at the same time gripping the hook of the weighing scale with the other hand(probably a left hand) steadily before finally placing the bucket handle on the hook.

(b) **Double Handle & Side technique** – this weighing technique as practiced by some operatives is done by grabbing the weighing bucket with both hands by the side (after applying some appropriate strength because of the weight of the solid minerals ores inside the bucket) raising the bucket and then locating the hook and hang the bucket on it and gently freeing the hands from the bucket.

Finally, the bucket is gently and steadily held to prevent swinging from side to side because it is capable of distorting the reading and gives a false and inaccurate reading. Once the bucket is still, the reading will be observed closely, and noted before recording it in the company's notepad. Lastly, the bucket will then be dropped from the weighing scale.

(G) **Bagging:**

the method of bagging for Processed Tin and Columbite is the same, with only exception in their quantity. The material used in bagging both Tin and Columbite whether in the final processed state or in the raw Tailing form is the normal salt bag.

Exceptions: Tin being the relatively heavier ore of the two is bagged in a standard weight of 70lbs pounds weight. This means that one (1) bag of processed Tin gives 70pounds weight. After bagging, the mouth of the bag is tied firmly in place using nylon ropes. Columbite on the other hand is bagged in a standard weight of 80lbs pounds weight. This also means that a standard bag of processed Columbite will give 80pounds weight.

After the bagging process, most mining companies will go ahead to demarcate the bags with ink marker writing on the bags exterior surface to include type of mineral, grade and serial numbers. This is indispensable for many reasons amongst which are

* to prevent confusion that may arise as a result of mixing different minerals ores together

* to prevent lost and theft

* for ease of handling

* for speed of transaction etc.

The same insignias stamped on the exterior surface of the bags will assist the loading and arrangement of the bags inside conveying mediums such as train wagons and Trailer trucks and taken to exporting points.

[4] GRADE BOOSTING TECHNIQUES

(A) **Roasting:**

The process of boosting or upgrading the initial grade of mineral ores from low grade to high grade through extensive thermal action which results in burning and melting off of impurities from the columbite in a local furnace is referred to as "roasting". This activity concerns the transformation or conversion of low grade mineral ore to a high grade one. This is unique because (I) firstly, not all miners are aware of its existence except those that came across it one time or the other, and are engage in the practice.

(II) Secondly, it is a cost effective engagement even though a rewarding one at the same time. It is a very rewarding venture if a miner can afford to set it rolling. The type of miners who engage in this activity are the directors who owned their tin sheds or other top level miners who can afford to buy large quantity of low grade columbite and monoxide and also provide the required facility. Mineral ore roasting has the capacity to provide 5 to 8 times profit or return on initial capital invested in. Roasting could be carried out on some mineral ores, but the most prominent and tested ones are columbite and monoxide ores.

(i) **Roasting of low grade columbite ore;** Low grade columbite ore, which are found from the following processes are utilized in the roasting process

(a) Low grade columbite from side discs of magnetic separators who are drawn along side with iron as a result of excessive current or much lowered magnetic disc.

(b) Low grade columbite from buckets 3, 4, and 5, from the air floating machine during "free floating" of very rich columbite ore freshly dug from Loto or during free floating of previously roasted columbite ore. (c) "Natural" low grade columbite ore which are sold as low grade after much processing because they cannot pass to the high grade status without exhausting the whole of their quantity.

(d) Low grade columbite from the back pan of "Jirgi"

during washing of medium grade columbite to raise its grade.

All the above mentioned sources of low grade columbite ore are being acquired by the miner through procurement or accumulation over time to become many before roasting action could be taken on them. The low grade columbite grade that needs to be roasted falls in the lowest range of 20.6 – 21.7 in the Burreting scale. The lowest grade columbite ore will have to be mixed with the medium range columbite ore which fall in the range of 19.4 – 20.5 in the Burreting scale. The indispensable mixing of the lowest grade with the medium grade columbite ores will ensure a guaranteed, fast and easy transformation to high grade columbite after roasting. Roasting of low grade columbite and monoxide cannot take place without the following major and minor tools and equipments ready in place.

Local brick Kiln; referred to as "oven" this depends on what shape the miner wants it to be. Mostly, the local Kiln is a rectangular or square shaped brick building with a whole open façade. It has compartments each having a hole referred to as a "Pit". Some local Kilns have up to 6-8 Pits created in it. The whole Kiln is clad with a roof which could extend up to 2m away from the oven to the front as a porch, to protect against rain. Each oven pit is built to a height of 1m from the ground and +_1.5m square in top area; each oven has one round hole of about 35-40cm in diameter. The oven holes are spaced to be able to accommodate logs of wood. The ovens are built in series to allow for easy and quick observation and to provide wide and easy access working space.

Hollow steel cylindrical pipe; this is a hollow steel pipe which is 3-4mm thick with a diameter of 150-200mm and a length of 2.2m – 2.5m long. This steel pipe must be held with both hands during chocking.

Flat base iron rod; this iron rod is relatively heavy and fabricated for the purpose of roasting; it is between 0.8 - 1.2m high with a diameter of 40-50mm wide. The bottom end of this iron is 100-120mm in diameter and constructed round to provide a wider area for ramming or choking of the mixture to enable compaction. It can be held with one hand or both hands during choking for the purpose of compaction.

Logs of firewood; Firewood is more economical than any potential type of fuel to be used for roasting and it produces greater burning effect as the external burner for the mixture. Logs of firewood are needed because the process consumes a lot of firewood. If small stacks of firewood are used, it will burn rapidly without producing much effect and be more capital extensive, therefore to achieve high thermal energy target, logs of wood which are bigger must be used.

*Diesel fuel; Diesel is the internal burner that heats and burns the mixture of both low grade columbite and sawn dust together. Diesel is used for roasting because it burns deeply and consistently slowly rather than premium motor spirit which burns rapidly and randomly.

*Sawn wooden/particle dust, this is the end product of timber sawing from timber cutting and sizing mills. It is the bye particles of wood cutting in some large carpenter shop. Sawn dust is usually gathered and sold in big sacks. Sawn dust is the preferred material for roasting because its small size is able to mix properly with the low grade columbite and more so it mixes very well with the liquid diesel.

*Long handled rectangular shovels; this is a rectangular shaped shovel which is made of steel pan measuring 175mm-190mm wide by 27mm-30mmm long.

It is narrow so that it can be able to enter into the pit hole to extract the columbite ores that are well roasted and fallen. The long handle of the shovel, which is up to 1.7m-2.3m long, is made from hard timber so that the workers can successfully remove roasted columbite ores from the oven pits that have fallen down from a distance without having to come close to the intense heat of the furnace.

***Brooms;** good strong brooms are very useful in columbite ore roasting. The brooms are used to sweep the already roasted columbite ore that have cooled off together in a heap since the floor is smoothly rendered. The brooms are used to sweep both the hot and cold roasted columbite ores that is why most of the brooms found here have their top blackened and burnt short because of the heat involved.

Roasting Process;

first of all, the columbite ore to be roasted are all gathered together and given a befitting mix. That means all the low grade columbite ore are mixed together with the medium grade columbite ores and probably half bag of a high grade columbite of say, between 18.2-18.4 in the Burreting scale. The next action is to spread the already mixed columbite ore widely on the ground and then add sawn dust and liquid gasoline or Diesel oil together and remix thoroughly. This mixture of liquid Diesel fuel, sawn dust and low grade columbite ore is a semi- paste being wet by the gasoline.

The mixture is then parked and choked together into pits. During choking, one of the workers supply and feeds the mixture into the pit while the other better experience worker positioned himself in the pit hole and holds the cylindrical hollow steel pipe in position at the center of the 35-40mm round pit hole with the pipe touching the bottom of the pit. When the mixture is poured all around the steel pipe, the experience worker holds the cylindrical steel pipe with one hand while at the same time use his other hand to grab the flat base steel rod and use it to ram the mixture continuously changing position round-about until compaction is achieved, this process is referred to as "choking". Once the 1m high pit is filled with the mixture and successfully compacted, the worker will remove the round steel pipe from the pit- this creates a round void space at the center of the pit equal to the diameter of the cylindrical hollow steel pipe. The void created allows for oxidation during combustion. The workers then move to the next pit carrying their tools along and repeat the same process for the subsequent pits until they were able to complete the number of pits available or intended to be used.

Once the choking process is carried out, the whole oven

is set ablaze and continues to burn. When matured or well roasted, the burnt columbite ore continue to fall down the pit hole and it is removed with the long shovel in order not to block the hole from getting oxidized. Maturity time is between 3-5hours depending on so many factors amongst which are viz

(I) the availability and quality of firewood- in this case well dried firewood.

(II) The nature of low grade used-whether it has much or little impurities.

(III) The quality of mixing done for the three main ingredients- that is low grade columbite, Diesel and sawn dust.

(IV) The richness of the mixed constituents-good and well dried sawn dust, unadulterated Diesel fuel.

(V) The quality of choking undertaken because if not properly choked, the mixture will "collapse down" and need to be choked again consuming time, energy and resources. If well and properly roasted, the irons in the low grade columbite are melted and become stiff while the monoxide content and zircon are total burnt to ashes.

(ii) **Roasting of Monoxide ore;** monoxide is a heavy and reddish color mineral ore. It is a combination of the following named ores in certain defined proportion

(a) good grade columbite

(b) very small quantity of iron

(c) small amount of Zircon and sand and

(d) large quantity of reddish-brown and rubber-like particles.

The reddish-brown and rubber-like particles give monoxide its color outlook. Monoxide is derived from the following process

(I) The final air floating of columbite ore using hand

brush to remove monoxide and other impurities..

(II) During final "passing" of columbite ore in the magnetic separator machine where columbite is drawn to the side discs leaving most of the monoxides to slip pass on the magnetic conveyor belt to the front pan along with tin ore, zircon and sand.

(III) During "passing" of (mostly washed) "over-belt" material to extract monoxide to the side discs of magnetic separator machine by dropping down the magnetic discs and increasing their current flow.

Roasting of Monoxide ore is very similar to roasting of low grade columbite using the same tools, equipments and thermal energy medium. The major difference between the two comes in the aspect of the Kiln. While low grade columbite prefers the kiln to achieve 100% success, monoxide can be roasted on flat thick metal plate, on a smooth rendered floor or on simply build roasting place.

(B) **Combinations/Blends;**

in mining of tin and columbite on the Plateau, combinations or blends means mixing of two or more mineral ores together to achieve the uplifting of one or the other in the mixed ores sample for economic reasons. The final outcome is directly dependent on the relative density of the booster ore. In this scenario, one of the mineral ore is obviously more economically superior or valuable than the other(s). This activity seems to achieve advantage in the Burreting aspect of mineral ore processing but not on the analysis aspect. The superior mineral ore is usually higher in specific gravity than the inferior one which needs boosting. Sometimes, the superior mineral ore only enjoys the advantage of higher relative density or mass over the inferior mineral ore (because it needs boosting) which sometimes enjoys the advantage of higher value rating or price.

The "booster" mineral ore is always heavier with greater atomic mass than the "boosted" which has lower atomic mass but could be more or less valuable.

Example:

Wolframite is heavier than a high grade columbite, with a higher specific gravity and enjoys 16-17 class in the Burreting grade, but is far less valuable than even a middle grade columbite reading 19.4 in the Burreting grade.

Some booster mineral ores are heavier and economically more valuable than the boosted mineral ores. The following are some ways to boost the grade of some mineral ores through mixing them together referred to as "concoction" in mining circles.

Boosting Combinations

(i) **Boosting low grade columbite ore with tin ore**;

"Booster mineral ore" (Tin ore) + "Boosted mineral ore" (low grade columbite ore) = medium or high grade Columbite.

A good grade tin ore gives a grade of 17-18 when Buretted. At the time when columbite ore is more expensive than Tin, and Tin having more specific gravity than columbite will be used.

A bag of tin ore will be mixed with several bags of low grade columbite say, 5 to 6 bags of 20.0-20.3 to produce high grade columbite bags of 18.3-18.5 all things being equal. The tin ore mineral will be well dispersed in the ensuing new bags of columbite that it could barely be detected even by acid and powder burning.

The miner will earn far more income by selling many

bags of high grade columbite even though; it is not an encouraging or legal practice.

(ii) **Boosting low grade columbite ore with wolframite ore;**

"Booster mineral ore" (Wolframite ore) + "Boosted mineral ore" (low grade columbite ore) = medium or high grade Columbite.

A good grade Wolframite ore gives a grade range of 15-17 when Buretted. Columbite ore has always been more expensive than Wolframite ore, but Wolframite having more specific gravity than columbite ore will be used. One (1) bag of Wolframite ore when mixed with several bags of low grade columbite say, 6 - 8 bags of 19.8-20.3 will produce high grade columbite bags of 18.1-18.5 all things being equal.

The wolframite ore will be spread thin in the new bags of columbite after proper mixing, making it difficult to detect the Wolframite using appropriate test. High grade columbite is usually sparkling and the presence of Wolframite will give it additional luster and sparkling look.

The miner(s) income will be multiplied several times when this is achieved, even though; it is not an encouraging or legal practice.

(iii) **Boosting low grade columbite ore with roasted monoxide;**

"Booster mineral ore" (roasted monoxide) + "Boosted mineral ore" (low grade columbite ore) = medium or high grade Columbite.

Monoxide ore on its own is a useless mineral ore but when roasted and dressed, it becomes a utility mineral because it can be sold on its own or it can be an additive ingredient for boosting the grades of lower inferior mineral ores. Good grade roasted monoxide ore will give a grade of 17.8-18.0 when Buretted. This range suggests that the principal constituent ore in a roasted monoxide ore is high grade columbite. High grade columbite is denser and therefore far heavier than any low grade columbite.

If monoxide is roasted and processed, one (1) bag of it can be added to 4-5bags of low grade columbite of say, 19.8-20.1, it will boost its capacity and convert it to high grade columbite, all things being equal. Here, there is no need to conceal its content because it is columbite itself. A miner will earn multiple returns on investment by selling many bags of high grade columbite boosted by roasted monoxide ore.

(iv) **Boosting low grade columbite ore with Lead ore;**
"Booster mineral ore" (Lead) + "Boosted mineral ore" (low grade columbite ore) = medium or high grade columbite.

A good grade Lead ore gives 17.6-18 when Buretted. Lead has less economic value at the time when columbite ore is more expensive, but Lead has higher specific gravity than columbite therefore it qualifies to serve as a good additive to boost low grade columbite to a new higher grade.

A bag of Lead ore when mixed with several bags of low grade columbite say, 4 to 5 bags of 19.5-20.3 will produce bags of 18.3-18.5 grade columbite, all things being equal. Lead ore which is dark blue black color will be well blended with the resulting new bags of columbite. A Lead test can easily reveal this concoction. Any miner that has this blend will earn more profit from selling higher grade columbite than selling low grade columbite or more so than an abandoned lead ore.

(v) **Boosting low grade columbite ore with Low grade Tantalite ore;**

"Booster mineral ore" (Low grade Tantalite) + "Boosted mineral ore" (low grade columbite ore) = medium or high grade Columbite.

A high grade tantalite ore gives a grade of 15.1-17 when Buretted. Tantalite has more specific gravity than either of columbite or tin. Tantalite is a coarse, dull mineral ore which comes in sizable grains and needs to be pounded or crushed. At the time when tantalite is very expensive, blending high grade Tantalite with columbite or Tin will be a waste of resources.

But a low grade tantalite will give 17.4-18.2 grades when buretted, this makes it suitable for blending with low grade columbite or tin to boost their weight s and therefore their grades.

If a bag of low grade tantalite ore is mixed with several

bags of low grade columbite or tin say, 7 to 8 bags of 19.7-20.2 to produce high grade columbite, it will produce bags of 18.1 -18.3 columbite or tin ores, all things being in order. The tantalite ore will be well spread in thin quantities in the mixture.

The miner here will earn far more income by selling many bags of high grade columbite than selling one bag of low grade tantalite. Even though; it is not an encouraging or legal practice.

[5] FINAL PRODUCT:

Mineral elements such as Tin (Sn), Niobium (Nb), and Tantalum (Ta) in their pure form emerged finally from the processing of its respective naturally occurring mineral ores or chemical compounds such as Cassiterite (SnO_2), Columbite (FeMn) Nb_2O_6 and Tantalite (FeMn) Ta2O6.

***Tin (Sn)** as a final product is derived from Cassiterite ore which is a raw mineral ore that has associated minerals such as tantalum, zircon, columbite, iron, manganese, titanium.

***Niobium (Nb)** otherwise generally referred to as 'columbite" in tin sheds and mining circles, is the final product derived from the processing of Columbite ore (FeMn) Nb_2O_6. Minerals associated with columbite ore concentration are tantalum, Cassiterite, wolfromite, iron and titanium.

***Tantalum (Ta)** is the final product derived from the processing of its raw mineral ore referred to as Tantalite (FeMn) Ta_2O_6 which has associated minerals like iron, manganese, tin, columbite, titanium.

(A) **Grading**;

this refers to the classification of final or finished product (mineral elements) from processing of raw mineral ore or tailing, after undergoing various processing stages. Grading is done to help define the quality of a final processed mineral element to determine its economic value or market price. Normally, grading classification of tin, niobium and tantalum is verified through volumetric analysis or burreting and through XRF analysis. When using the energy dispersive x-ray fluorescence analysis, grading is normally displayed or output in percentage of total scan of existing chemical compound or mineral ores in any given sample. Here, when a final processed mineral element like tin is analyzed using ED-XRF, it gives out grading in percentage of existing constituent elements in the sample 79.4 % Sn; 8.9 % Fe; 2.5 % Ti; 1.5% Ta; 4.8 % Nb; and 2.9 % W. When using volumetric analysis or burreting, grading is normally classified and presented based on volume differential readings from graduations on the pipette.

Burreting is carried out on finally processed mineral elements where the ensuing result is influenced by their individual weights.

It was observed that the individual weight or specific gravity of mineral elements has relationship with their atomic mass, that is why iron (Fe) which has 26 atomic mass is lighter than Niobium (Nb) which has 41 atomic mass higher than that of iron, it equally follows that tantalum (Ta) with atomic mass of 73 is heavier than tin (Sn) which has 50 atomic mass.

The above fact is substantiated as observed from consistent range of burette results on individual mineral elements where iron (Fe) could give a burette result range of 25.0-22.0, Niobium (Nb) could give a burette result range of 21.5-18.0, tin (Sn) could give a burette result range of 18.0-17.5 and tantalum (Ta) could give a burette result range of 17.5-15.3.

(B) **Storing;**

Finally processed and partially processed mineral elements are stored as well dried products preparatory to disposal or further dressing. For finally processed high grade mineral elements, they are either sold to a bigger mining company or they are exported by a big mining company. The following are the two main types of storages

(i) Internal storage; this is the safe keeping of mineral products purchased by a mining company whether high grade, medium grade or low grades for disposal (high grades) or for future reference (medium and low grades). Storing of mineral elements is easy and it's independent of controlled environment no matter how long it takes to store. This means that it is non-perishable and it doesn't have to be stored on certain preservation conditions like some products that need to be freezed, while some needs to be air-conditioned while others needed to be warmed. Some years ago, most mining companies prefer to secure their finished products in a store referred to as "strong rooms". The strong room is usually wide and higher than the normal building height of 3m. Some big companies that trade in many high grade materials have their strong rooms constructed with double walls. Such security stores have tiny and high positioned windows and double steel security entrance doors with multiple iron mongeries.

The high grade products bought by the company are always stored at the inner spaces far away from the entrance door. They are contained in small bags and taken to the store for safe keeping. The interior of the stores is well arranged based on space allocation for specific type of mineral elements. Tin, Niobium, Tantalum and other mineral elements are allocated adjacent or opposite space for storage in the strong room. Written signs are posted throughout the store, high on the wall close to the location of specific stored bags of mineral elements stating the type of minerals. The bags of mineral elements are placed vertically one on top of the other and side by side in horizontal strata after placing the initial ones on a prepared fabricated steel or wooden platform covered with hard board. This platform which measures 40-50mm high from the floor level will prevent much cold and damp that could only stiffen the bags but can never damage the constituent minerals inside the bags.

Low grade minerals bought by the company are always stored at one side of the allocated space and demarcated so. These are always stored at the outer space of the store room for ease of parking back to the tin shed for further processing. For low grades, they are always stored with a paper inside the bags which have writings stating the type of mineral ore, the grade of the mineral ore and the weighted quantity of the mineral ore.

(ii) External storage;

(a) **Raw mineral ore Storage**: this is a type of warehouse constructed some distance away from excavation and paddock areas and more closer to the processing zone. The reason for this storage is to safe keep the raw tailing ores against weather elements like rainfall and against theft. The already excavated ores must be processed while dry. The ores are normally bagged in sacks and well arranged in vertical piles kept in the warehouse and properly allocated and arranged to demarcate gap for ownership. Each miner distinguishes his property from others by using ink markers or Ball pens. Sometimes, the Storage or warehouse is filled up to capacity. At the peak of tin and columbite mining activities on the Jos Plateau, so many miners have brought their raw tailings or mineral ores from far and near neighboring local government areas where the minerals are found, dug and transported to the raw minerals ore stores to await machine processing.

At the peak of Tantalite quarrying, some miners brought mineral ores from as far as Nassarawa, Zamfara, Ibadan and other distant states to Plateau state to access machine processing, analysis and sale. This led to queuing at the few storage facilities and shortage of storage space in the raw mineral ore stores thereby necessitating space allocation.

(b) **Processed mineral product Storage**:

this type of storage only concerns high grade mineral products. It is carried out external to the tin shed and far away from it. This type of storage is undertaken by a big mining company or contracted by the mining company to a conglomerate warehouse company like John Holt etc to store the products prior to transportation to sea port for export.

Processed high grade mineral elements could also be stored to an external storage facility by a middleman or an intermediary who procure from a mining company to export and sell overseas. In this case, such external storage facilities are very large warehouse building structures that are capable of containing hundreds of thousands of bags of processed mineral elements. Warehouse buildings are very tall buildings reaching up to 4.3m high and more or less 27-40m long and sometimes reaching 15-27m wide. Large warehouses have functional managers and many warehouses allow entrance, movement and exit of big haulage and conveyance vehicles. Warehouses are a well secured and well guarded premise that is why most shipping agencies take advantage of its services. Likewise, most of the big warehouses have their own transportation vehicles.

(C) **Transportation and recommended vehicles;**

(i) **Transportation;** transporting raw mineral ore or tailings and the processed mineral elements is achieved through land vehicles. Transportation is significant because it is involved throughout the value chain of solid minerals mining sub sector on the Plateau. Solid mineral is so heavy that it needs strong vehicles to do the conveyance. Transportation could be from the lotto or paddock to the raw mineral ore storage and from the raw mineral storage to the tin shed for processing. Transportation is also involved when transporting finally processed mineral elements to selling destination or to a warehouse for storage pending exportation. Waste products from tailing processing are also transported to dumping sites some as far as twenty kilometers away from the tin shed.

Some miners have alternative drying places other than the tin shed drying space which is always crowded, in that wise they have to transport the raw mineral ore or tailing to and fro the drying place at their own inclusive cost. This same cost will be built in their initial expenditure. The following are the various transportation channels in the solid mineral mining on the Plateau viz

* transporting mineral ore or tailings from lotto or paddock to raw mineral ore store for safekeeping prior to movement to tin shed for further processing.

* transporting mineral ore or tailings from raw mineral ore store to tin shed for washing and drying.

* transporting of mineral ore or tailings from tin shed store to drying place.

* transporting mineral ore or tailings from surface or

underground dump sites to the tin shed for washing and further machine processing.

* transporting mineral ore or tailing from drying place to tin shed for processing.

* transporting waste products from tin shed to dumping sites.

* transporting finally processed mineral products from tin shed to other tin sheds for selling or further dressing.

* transporting finally processed mineral products from tin shed to warehouse for safe keeping.

* transporting finally processed mineral products from warehouse to shipping port.

* transporting finally processed mineral products from ports to overseas.

Some distances are near enough to be covered by human effort rather than mechanical effort. Conveyance from one point of operation to another in a tin shed may not necessarily require the use of vehicles even if it is the better alternative in saving time there. At the peak of tin shed operation, there could be up to 60-150 people in and around the tin shed engaged in one form of activity or the other. This is not to mention the piles of tailings that littered the surrounding environment, the presence of machines in strategic locations around the tin shed. All these when found in one tin shed at a time could prove a barrier for free moving vehicles. This means that a vehicle cannot maneuver its way around obstacles on the tin shed ground to convey mineral ores from the store to the drying place or from drying place to the machine room; except in initially well planned tin sheds; this will require manual labor to get the job done.

(ii) **Recommended mining vehicles**;

(1)**vehicles for conveying mineral ores and miners;** the recommended utility vehicles for transportation of products and personnel from one place to another are the Land Rover four wheel drive with auxiliary gear utility vehicle or any stronger updated or alternative model, the Toyota Hilux double cabin pick-up four wheel drive vehicle and the Pajero Land cruiser utility vehicle or any stronger updated or alternative model.

(2) vehicles for conveying average(50 to 120) bags of excavated mineral ores and miners; the recommended vehicle for transporting mineral ores in bags and at the same time transporting miners who owned the mineral ores is the high body and fast moving J-5(as it is popularly referred to) or any similar updated or alternative model.

The Chevrolet wagon, an American van model vehicle which has a wide interior void space of up to 7.56 square with double back doors and sliding side door which is capable of conveying sizable number of bags of minerals ores and miners at the same time from excavation zones in the city outskirts or hinterlands and villages to storage or processing zone.

(3) **vehicles for conveying large quantity of tailings;** Tipper 8-10 tires diesel engine trucks are strongly recommended for conveyance of large volumes of raw mineral ore or tailing from excavation sites like lotto or paddocks or surface or underground dumps to tin sheds or dumping sites. This vehicle can carry up to 5-11 cubic meters of mineral ores or tailings at a time. It carries mineral ores in bags and those that are not bagged. For mineral ores not bagged, it deposits it by mechanical movement through inclined lift and drop referred to as "tipping" to the designated position.

(4) **Long distance Vehicles for conveying large quantity of processed mineral products;** the Trailer or mark long transport vehicles fall into this category of

transportation mediums. These types of vehicles use diesel engines and are suitable for long distance journeys with expansive carriage space for conveying goods and miners at the same time. Vehicles in this category can carry up to 500-800 bags of 70-80kgs of processed mineral products successfully to warehouse or to seaports several hundreds of kilometers away. The trailer or Mark long vehicles can also be used to carry large quantities of good grade mineral ores either in bags or as free tailings or both at the same time because of the available space in the vehicle and its strength capacity.

(5) **Goods transport trains;** for conveying super large mineral ores, goods transport train wagons are better recommended. In the early days of mining, the colonial masters built the railway grid system linking productive regions of raw economic materials in through the hinterlands to transport economic produce and solid

minerals with direct route to the sea port at Lagos for express export to smelting factories in the UK. Tin, Niobium and other solid minerals were mass transported in great tonnage through the railway system in goods wagons.

REFERENCES

[1] Bradshaw, M. J. (2005). Population, Resources, Development and the Environment. In: Daniels, P. *et al*, eds. An Introduction to Human Geography: Issues for the 21st Century. 2nd ed. (Section 2).

[2] Bridge, G. (2008). Economic Geography: Natural Resources. In: International Encyclopaedia of Human Geography. Eds. Kitchin and Thrift. Elsevier.

[3] Gyang, J. D., Nanle, N., Chollom, S. G. (2010). An Overview of Mineral Resources Development in Nigeria: Problems and Prospects. *Continental J.*

Sustainable Development, 1, 23-31.

[4] Solomon, M. H. (2000). Growth and Diversification in Mineral Economies: Planning and Incentives for Diversification. South Africa: United Nations Centre for Trade and Development.

[5] UNCTAD (United Nations Conference on Trade and Development), (2007). World Investment Report 2007: Transnational Corporations, Extractive Industries and Development. New York and Geneva: United Nations.

[6] Okeke, C. N. (2008). The Second Scramble for Africa's oil and Mineral Resources: Blessing or Curse? *TheInternational Lawyer*, 42 (1), 193-210.

[7] Mahtani, D. (2008). The New Scramble for Africa's Resources. Financial Times Special Report, 28 Jan. 1-6.

[8] Twerefou, D. K. (2009). Mineral Exploitation, Environmental Sustainability and Sustainable Development in EAC, SADC and ECOWAS Regions. *African Trade and Policy Centre Work in Progress*, 79. Economic Commission for Africa.

[9] Morgan, P. G. (2002). Mineral title Management - the key to Attracting Foreign Mining Investment in Developing Countries? Trans. Instn Min. Metall. (Sect. B: Appl. earth sci.), B165-B170.

[10] Girones, E. O., Pugachevsky, A. and Walser, G. (2009). Mineral Rights Cadastre: Promoting Transparent Access to Mineral Resources. Extractive Industries for Development Series #4 June 2009. Washington DC: The World Bank.

[11] Obaje, N. (2009). *Geology and Mineral Resources of Nigeria*. London: Springer (Chapter 1).

[12] Cowie, A. (2010): Tin, an Overlooked Commodity. The Market Oracle, Aug 19, 2010

http://www.MarketOracle.co.uk

—

[13] Davenport, J. (2010). Nigeria Aiming to Grow Mining's GDP Contribution to 15% by 2015. Mining Weekly, March 15, 2010.

[14] Pastor, J. and Ogezi, A.E., 1986, New evidence of cassiterite bearing Precambrian basement rocks of the Jos Plateau, Nigeria – the Gurum case study, Mineralium Deposita, Vol. 21, No. 1, pp 81-83.

[15] McAllister, M., Scoble, M. & Veiga, M. (2001). Mining with Communities. *Nat. Res. Forum*, 25, 191-202. Morgan, P. G. (2002). Mineral title

Management - the key to Attracting Foreign Mining Investment in Developing Countries? Trans. Instn Min. Metall. (Sect. B: Appl. earth sci.), B165-B170.

[16] Metallic Minerals, October2, 2010. http://www.onlinenigeria.com/minerals/?blurb=517 .

[17] Funtua, I.I., Idris, Y., Oyewale, A.O., Umar, I.M. and Elegba, S.B., 1997, Determination of Tin in cassiterite ores and tailings by 241Am source X-ray Flourescence Spectrometry, Appl. Radiat. Isot. Vol. 48. No.1, pp 103-104.

[18] Evans, J.R. and Jackson, J.C., 1989, Determination of tin in silicate rocks by energy dispersive X-ray fluorescence spectrometry. X – Ray Spectrom. 18,

pp 139.

[19] Tin Investing News, September 30, 2010, Tin Advances on supply threats. http://tininvestingnews.com/483- tin-advances-on-supply-threats.html.

[20] Abubakre, O.K., Sule, Y.O. and Muriana, R.A., 2009, Exploring the Potentials of tailings of Bukuru cassiterite Deposit for the Production of Iron ore Pellets, JMMCE, Vol. 8, No. 5, pp 359-366.

[21] Xiaohui W., Shili Z., Hongbin X. and Yi Z., 2009, Leaching of niobium and tantalum from a low – grade ore using a KOH roast-water leach system, Hydrometallurgy, Vol. 98, Issues3-4, pp 219-223.

[22] Ogwuegbu M., Onyedika G., Jiann-Yang H, Ayuk A., Peng Z, Li B., Ejike, E.N.O. and Matt A. (2011)

Mineralogical Characterization of Kuru Cassiterite Ore by SEM-EDS, XRD and ICP Techniques. *Journal of Minerals & Materials Characterization & Engineering,* Vol. 10, No.9, pp.855-863, 2011.

[23] Ofor O., 1992, Extraction of tin from cassiterite tailing by leaching – electrowinning process, Journal of Mining and Geology, Vol. 28, No. 1, pp 125-128.

[24] Adams, H., 2001, Columbia Encyclopedia, Sixth Edition, available at www.bartleby.com

[25] Calvert, J.B., 2000, Tin 4, available at www.du.edu/~jcalvert/phys/tin.htm

[26] Natasha, C., 2002, Tin mining of the Jos plateau, available at www.uni.edu/gai/nigeria/lessons/tin-mining.htm

[27] Gibson, O., 2002, "Tin Smelting in Nigeria – the Challenge of our time." Proceedings of the 19th Annual Conference/AGM of the Nigerian Metallurgical Society, pp. 47 – 60.

[28] Guanzhou, Q., Tao, J., Zhucheng H., Deqing, Z., and Xiaohui, F., 2002, "Characterization of Preparing Cold Bonded pellets for Direct Reduction using an Organic Binder." ISIJ International Journal, Vol. 4, pp. 20 – 25.

[29] Kurt, M., 1980, Pelletizing of Iron Ore, Springer – Verlag Berlin Heidelberg, New York.

[30] Rumpf, H., 1962, "The Strength of Granules and Agglomerates." Agglomeration (ed. W.A. Knepper), New York.

[31] Tohidi, N., and Rames, V., 1997, "Preparation of Iron and Steel Burden." Seminar paper presented at Tehran University.

[32] Kasai A., Murayama T., and Ono Y., 1993, "Measurement of Effective Thermal Conductivity of Coke." ISIJ International, Vol. 33, pp. 697 – 702.

[33] Journal of Minerals & Materials Characterization & Engineering, Vol. 8, No.5, pp 359-366, 2009 jmmce.org Printed in the USA. All rights reserved

[34] Exploring the Potentials of Tailings of Bukuru Cassiterite Deposit for the Production of Iron Ore Pellets O. K. Abubakre*1 , Y. O. Sule2 and R. A. Muriana

1Department of Mechanical Engineering Federal University of Technology, Minna, Nigeria 2 Raw Materials Research & Development Council, Abuja, Nigeria.

[35] American Journal of Mining and Metallurgy. **2017**, 4(1), 51-61. DOI: 10.12691/ajmm-4-1-5

www.ingramcontent.com/pod-product-compliance
Lightning Source LLC
Chambersburg PA
CBHW021817170526
45157CB00007B/2618